D0897867

HIGH TECHNOLOGY INDUSTRY AND INNOVATIVE ENVIRONMENTS:
The European Experience

HIGH TECHNOLOGY INDUSTRY AND INNOVATIVE ENVIRONMENTS:

The European Experience

Edited by
PHILIPPE AYDALOT and DAVID KEEBLE

ROUTLEDGE
London and New York

First published in 1988 by
Routledge
11 New Fetter Lane, London EC4P 4EE

Published in the USA by
Routledge
in association with Routledge, Chapman & Hall, Inc.
29 West 35th Street, New York NY 10001

Printed in Great Britain
by Billing & Sons Limited, Worcester.

British Library Cataloguing in Publication Data

High technology industry and innovative
environments: the European experience.
1. High technology industries — Europe
I. Aydalot, Philippe II. Keeble, David
338.4′76′094 HC240

ISBN 0-415-00414-4

Library of Congress Cataloging-in-Publication Data

ISBN 0-415-00414-4

Contents

Contents

Figures

Contributors

Mateo Alaluf, Institut de Sociologie, Free University of Brussels, Belgium

Philippe Aydalot, Centre Economie-Espace-Environnement, University of Paris 1 Panthéon-Sorbonne, France

Roberto Camagni, University of Padua and University Luigi Bocconi, Milan, Italy

Ewa Glugiewicz, Institut Planification Academie Economique, University of Poznan, Poland

Bohdan Gruchman, Institut Planification Academie Economique, University of Poznan, Poland

David Keeble, Department of Geography, University of Cambridge, United Kingdom

Guy Loinger, Centre d'Etudes des Systèmes et des Technologies Avancées, France

Dennis Maillat, Institut de Recherches Economiques et Régionales, University of Neuchâtel, Switzerland

Jean-Claude Perrin, Centre Economie Régionale, University of Aix-Marseille III, France

Veronique Peyrache, Centre Economie-Espace-Environnement, University of Paris 1 Panthéon Sorbonne, France

Claude Pottier, LAREA/CEREM, University of Paris X, France

Remigio Ratti, Ufficio delle Ricerche Economiche, Bellinzona, Switzerland

Adinda Vanheerswynghels, Institut de Sociologie, Free University of Brussels, Belgium

Jean-Yves Vasserot, Institut de Recherches Economiques et Régionales, University of Neuchâtel, Switzerland

Preface

This book incorporates a series of theoretical and empirical assessments of the impact on local and regional development in Europe of technological innovation and high-technology industry. It is the product of recent work by members of the **Groupe de Recherche Européen sur les Milieux Innovateurs (GREMI),** comprising a number of European researchers from different countries whose work under the leadership of Professor Philippe Aydalot has already led to the publication in 1986 of the volume Milieux Innovateurs en Europe (GREMI, Paris, 361 pp). The prime focus of the Group's research activity has been on contemporary processes of technological change in European manufacturing industry and the ways in which these are shaped by, and impact on, local environments, their existing firms, workforces, and social and economic structures. The GREMI has sought not only to understand in greater depth the interactions between local industrial environments and technological change, but also to assist in policy formulation for local economic regeneration and development.

Most of the contributions in this book were also presented at the 25th annual conference of the **Association de Science Régionale de Langue Française,** which was held at the Ministry of Research and Technology in Paris, September 1-3, 1986. This conference, on the theme of Technologies Nouvelles et Développement Régional, was also organised by Professor Aydalot, working through the Centre Economie-Espace-Environnement of the University of Paris 1 Panthéon-Sorbonne, the research centre of which he was founder and director. Originally written in French, the conference papers selected by Professor Aydalot and GREMI for inclusion in the present volume have been translated into English in order to acquaint as wide an audience as possible, both inside and outside Europe, with the Group's findings and research insights.

This book is thus very substantially the initiative of **Professor Philippe Aydalot,** as inspiration, founder member and chairman of GREMI since its inception in 1984, organiser and chief architect of the 1986 Paris ASRLF conference, and French editor of the papers selected for publication here. It is therefore with enormous sadness that his co-editor and fellow GREMI researchers record Philippe

Preface

Aydalot's sudden death on April 5, 1987, at the age of only 48 years, while in the very process of editing this volume. His death, as one of France's and Europe's most productive, enterprising and stimulating researchers on contemporary spatial industrial change, is a tragic loss to the European research community, as of course also to his widow and daughter. His fellow researchers and co-editor wish to place on record their profound debt - in ideas, academic stimulus and personal friendship - to Philippe Aydalot, and dedicate this book to his memory, as a tribute from all of us to his scholarship and outstanding contribution to European regional industrial analysis.

Chapter 1

High-Technology Industry and Innovative Environments In Europe: An Overview

Philippe Aydalot and David Keeble

DEFINITION AND CONCEPTUALISATION OF HIGH-TECHNOLOGY ACTIVITY

Amongst the various factors which may permit an understanding of current industrial and economic change in Europe, the role of new technologies is one of the most important. New technologies are one of the most visible aspects of the dramatic transformation of European society and economy which appears to be in progress at the end of the 20th century, involving daily life, work relations, and all aspects of economic organisation. At the same time, major changes are occurring at the spatial level. Rural industrialization contrasts with largescale urban manufacturing decline, crisis in traditional industrial areas contrasts with new dynamism in once-backward regions, "snowbelt" de-industrialization contrasts with "sunbelt" industrial take-off. Recent reversals in migration flows, and shifts in the geography of unemployment, graphically illustrate the effects of these changes upon the spatial economic landscape. Such changes cannot however be understood without considering the disruptive and creative effects of new technologies upon firms, industries and regions.

These effects are most frequently associated, in popular impression if not academic analysis, with the development of socalled "high-technology industry". What however is high-technology industry? The growth in attempts at its definition only serves to highlight the lack of agreement on this issue. Is high technology the introduction of particular technical changes within the production process, or is it the evolution of a whole new "information technology" paradigm? (Castells, 1984: Freeman and Soete, 1984). There is often confusion between definition and classification (Haug, 1986). Although general ideas about new technologies may provide an appropriate starting-point, the stumbling-block of how to match specific indicators with a list of high-technology activities is quickly reached. The following represent some of the approaches or types of indicators which are present in the literature (Glasmeier, 1985a):

i) public opinion or the views of experts on the list of products or activities judged to be indicative of the new technologies. These are based largely on subjective value judgments.

ii) the rate of change in employment or output. The underlying idea here is that while old industries decline, new activities based on new technologies are likely to exhibit rapid growth in sales, output and employment. However, as workers such as Glasmeier (1985a) and Thompson (1987) point out, some traditional industries can perform better than those often considered as high-technology. Thus in the USA between 1965 and 1977, employment in the furniture industry grew faster than that in electronics (Glasmeier, 1985a).

iii) the ratio of R and D expenditure to sales may give some idea of an industry's capacity for technical evolution and its "information" content. However, sectors with a large volume of sales such as petroleum-refining are probably under-valued by this measure compared with younger sectors whose turnover is still small. The index thus perhaps measures rate of technical progress rather than the absolute level of technology incorporated into production: while all else being equal, a higher level of competition in a sector may also lead to a greater concentration on R and D.

iv) the proportion of technically-qualified workers in the total workforce. This frequently-used measure (Glasmeier, 1985b) has the advantage of facilitating intersectoral comparisons, but its links with a clear definition of high-technology activity remain to be considered.

Underlying these questions and uncertainties are many ambiguities due to the lack of a coherent attempt at definition. Among the ideas implicit in various lists of high-technology sectors, there are at least four underlying definitions of high-technology activity, none of which is fully satisfactory. These are:

i) high-technology as a young or new activity. Examples might be semi-conductors in the 1960s, and micro-computing at the end of the 1970s. The problem here however is that the majority of activities usually considered as high-technology are relatively old. Thus the telecommunications and aerospace industries first came into existence more than a century ago, while the electronics industry is now 40 years old. This perhaps explains the frequent failure of attempts to identify in such sectors less concentrated organizational structures focussed on small and medium-sized firms, of the type often supposed to characterize the early stages of development of new sectors (Markusen, 1985). Indeed, Glasmeier (1985a) and Swynguedow and Anderson (1986) have shown that high technology sectors are often more concentrated, in terms of domination by large firms, than the average.

ii) high-technology as an innovative industry. Do high-technology sectors innovate more than others? This is not certain. After all, both the automobile and steel industries have adopted many new innovations in recent years, scarcely less perhaps than the electronics sector. The latter does however have a much greater rate of product rather than process innovations, which is probably a more significant measure of high-technology activity. The problem then arises of classifying innovations as "process" or "product".

iii) high-technology as an industry whose products alter the behaviour of both individuals and groups in society, as well as economic practices. From this viewpoint, the electronics sector does appear to warrant definition as a high-technology activity because of its widespread current use and impact socially and technologically (computers, telecommunications...).

iv) high-technology as a "science-based" activity. The emphasis here is on the fact that some activities have high proportions of their workforce engaged in the production of applied knowledge. However, is knowledge content the key point here, or is it rather that certain activities exhibit in extreme form a functional division of labour in which relatively few routine workers, whose work has little skill content, are contrasted with many non-production staff, in whom virtually all the knowledge required is vested?

Thus, for example, it is generally agreed that the aerospace industry, a relatively old industry by date of origin, is a high-technology activity. Is this because it exhibits a high innovation rate, because it uses a high proportion of products developed out of recent, new technologies, because it frequently changes its products, or because it permits and creates new forms of social behaviour, such as the international movement of people? The lack of a clear definition perpetuates the underlying confusion over what should be classified as high technology. One of the consequences of this uncertainty is that a very broad definition is usually adopted, while indicators may be chosen less for what they really express than for the extent to which they fit a pre-established list of sectors.

In effect, two contrasting approaches exist to the study of high-technology industry. These need to be distinguished in order to understand the nature of the choice which has to be made. Emphasis can either be focussed on the special nature of new activities and the types of changes which they engender in society: or on the degree of rupture with previous technologies and forms of social and economic organisation which they cause. In the former case, the focus is on the nature of the change, in the latter, on the amount of change. Are the recent upheavals in Europe's industrial economies the expression, to greater or lesser degree, of the invasion of our societies by particular radically-new technologies and activities? Or are they to be viewed as the first phase of a longer-term cyclical process? Which of these two aspects, both of which do seem to be present, is the more important? If the key point is the initiation of the first phase of a new economic cycle, then comparison with previous periods of technological upheaval may be helpful. If however the key factor today is the specific and distinctive properties and characteristics of current new technologies, then a different approach is required.

In comparison with previous technological revolutions, that which we have been experiencing during the past ten years clearly possesses distinctive characteristics. While earlier revolutions were, for the most part, linked to the emergence of new forms of energy (the steam engine and the railway, the combustion engine and the car), this one is focussed on the production and supply of information in all its forms. It has very important non-material aspects, concerned less with products than with the quality of knowledge used to produce them, and the diffusion of knowledge. It is transforming an economy which was based on the production of material goods into a machine for the creation and transfer of information. The productive processes themselves depend upon techniques which rely upon information. It is therefore easy to argue from this the case for the originality and specificity of current new technology, as well as for its particularly powerful and widespread impact: all aspects of daily life, of social behaviour, and of production are being affected. The form of settlements and towns, the organisation of work and leisure time, the very nature of leisure itself, all increasingly bear the imprint of these new technologies.

There is an alternative way of interpreting the current situation. From this perspective, high-technology activity symbolises and causes the rupture or reversal of a previous long cycle of economic change, the obsolescence of traditional technologies and of the industries and regions which embodied them. It involves new patterns of investment, focussed on new firms, new qualifications and skills, and new labour markets. Since the beginnings of industrialization, it is possible to identify successive phases of sustained economic growth and crisis. The rhythm of economic development appears to conform to long waves of boom and slump, with many observers arguing that the 1980s mark the beginning of the fifth Kondratief cycle (Hall, 1981, 1984: Freeman, 1986). This has been preceded by others, each based on a new and different technology. The crisis of the 1970s bears comparison with that of the late 1920s and early 1930s, whilst the last two decades of the 19th century were equally characterised by severe economic depression. So from this perspective, it is not a question of an historically-unique change resulting from the development of a totally new technology which has divided European history into two periods - before and after the information technology revolution - but rather one of major periodic technological, economic, and indeed spatial, re-alignments.

From this viewpoint, it can be argued that notwithstanding the current importance of new information technologies, the steam engine at the beginning of the 19th century, the internal combustion engine at the beginning of the 20th century, and other major innovations have played an equally important role in past periods. This in turn suggests - and this is the crucial point - that more deeprooted economic, social and spatial processes predate the development of new information technology, which is no more than the catalyst or driving force for a new cycle. On this argument, the determinants of spatial organisation have as much to do with these wider underlying forces as with the particular nature of information technology per se.

Both these two interpretations are tenable. However, they do not result in the same list of technologically-advanced activities. In the first case, analysis focusses on micro-electronics, computers, and associated information technology industries and services. In the second case, however, all new technologies which have appeared in the last few decades, such as bio-engineering, laser technology and aerospace as well as micro-electronics and computers, must be included in the high-technology sector.

Equally, the first interpretation does not point to identification of high-technology activities on the basis of the level of research and development activity, or of new technology, involved. Rather, it suggests that they should be defined in terms of the degree of upheaval they engender in individual and collective behaviour, in the sphere of work as well as in daily life. In contrast, the second interpretation directs attention to establishing a list of the significant innovations which give birth to the new sectors and economic activities which come to replace those of earlier cycles.

The challenge presented by the new technologies to the spatial analyst stems from the fact that their development calls into question traditional ideas and theories of the location of industrial activity and regional economic structures. For many years, the dominant trend in Europe and Northern America seemed clearly to be one of spatial concentration. Explanations were developed around notions of the dominance of a leading sector or firm (Perroux, 1955), of the formation of industrial complexes (Czamanski, 1971), and even of the simple mechanism of attraction. Today, through the concept of synergy, there is new interest in the processes which may lead to firm clustering in geographic space.

However, since at least the early 1970s, the general emphasis has switched to the observation and explanation of spatial dispersion, now apparent in virtually all industrialized countries and at all spatial scales (Keeble, 1976: Keeble, Owens and Thompson, 1983). One fruitful approach to an understanding of this trend is that invoking the development of a new spatial division of labour (Aydalot, 1976, 1983: Hymer, 1979: Massey, 1979) as a consequence of the hierarchical dispersion of functions by large, multi-plant industrial companies. The growth in the importance of such organizations, at least until the 1970s, and their ever-widening geographical area of operations, has enabled them to exploit regional labour market variations through systems of functionally-specialized but spatially-dispersed branch establishments (Keeble, 1987, 6-9).

Attempts to render this approach more dynamic have focussed on how firm spatial organisation has changed with changing phases of a given long cycle. Vernon (1966), Aydalot (1976) and Markusen (1985) have each showed in turn how early stages of the product/profit cycle appear to be characterised by the co-existence of dispersed industrial (firm-size) structures and concentrated spatial structures. Growing product maturity leads to an increasing dominance of large firms and the creation of a spatial division of labour in the later stages of the cycle.

The attractiveness of this dynamic model of the structural and spatial organisation of production does not however solve all the problems of understanding contemporary spatial patterns of industrial change. Thus, for example, the production of integrated circuits, the most technologically-advanced and representative branch of the semiconductor industry, exemplifies a complex historic and spatial process of evolution, which reveals both a growing concentration around a major centre, Silicon Valley, and an increasing and contemporaneous spatial dispersion, the first semiconductor assembly line in South-East Asia being set up as long ago as 1962. Semiconductor plants can today be found in most American states and in a host of other countries, notably those of the Third World (Henderson and Scott, 1987).

As this example illustrates, problems of understanding industrial location have re-emerged in recent years through the observation of new industrial sectors which stem from advanced technologies. Nonetheless, it may be argued that studies which focus on the structure and organisation of industry in the context of the evolving spatial division of labour today represent a particularly fertile field of analysis, which permits a progressively-widening understanding of the spatial logic of firms and industries.

INNOVATION, NEW TECHNOLOGIES AND THE LOCAL ENVIRONMENT

If new technologies are renewing the dynamism of local or regional industrial environments, or creating new industrial concentrations in hitherto less-developed areas, what is the best angle of attack to analyse these changes? There would seem to be three possible approaches:

i) to begin with the firm or enterprise, to investigate its locational history, and hence develop a better understanding of the factors involved in the location of industrial activity which is based on new technologies.

ii) to begin with the technologies themselves, to assess their impact on regional development. Are they creating employment, and is this employment located evenly or unevenly across geographic space? Will they increase regional differentiation in economic development potential, will they reinforce or reduce regional disparities?

iii) to begin with the local environment. Where do innovations appear? Which local environments appear to be best in creating, diffusing, and generalising the use of technological innovations?

The enterprise-focussed approach highlights the fact that those new activities which are most influenced by and involved in technological innovation do not exhibit the same locational behaviour as older industries (Hall, 1981). This allows an understanding of territorial industrial restructuring under the impact of technological change, but from the locality's or region's perspective, it is an understanding "from the outside". The areas, the local communities and economies, which are chosen - or not chosen - as production locations by the firms involved tend to be viewed as passive objects of external decision-making. The spotlight is on the firm, as the dominant actor with a capacity for autonomous behaviour. With this approach, local environments tend to be considered only as passive suppliers of factors of location and production.

The technology-based approach considers the nature of new technologies as given and analyses their impact on territorial equilibrium. However, this kind of approach can appear too global, too generalized. Though very valuable from a regional policy perspective and useful in assessing the likely evolution of regional disparities, it tells us more about the outcomes than about the processes of change.

The local environment-based approach is arguably the most fruitful. Its central concern is to understand the firm in its local and regional context, and to ascertain what conditions external to the enterprise are necessary both for the creation of new firms and the adoption of innovations by existing ones. The firm, and the innovating firm, are not viewed as pre-existing in or separate from the local environment, but as being products of it. Local milieus are regarded as the nurseries, the incubators, of innovations and innovative firms. This approach implies that innovative behaviour is as much dependent on variables defined at the local and regional level as on national-scale influences. Access to technological know-how, the availability of local industrial linkages and inputs, the impact of close market proximity, the existence of a pool of qualified labour - these are the innovation factors which will determine areas of greater or lesser innovative activity within national space. New technologies are often developed or first applied by new firms and enterprises (Rothwell and Zegveld, 1982: Keeble and Kelly, 1986): and these are usually created within (and by ?) the local environment in which they first appear. In such cases the enterprise is not an independent agent freely choosing its location from a range of alternatives: it is rather a direct product of its own particular local environment.

This line of argument leads naturally to the hypothesis that it is often the local environment which is, in effect, the entrepreneur and innovator, rather than the firm. In this context, a major problem in innovation adoption is the cost of adopting new technology, which varies between localities. Overall, however, it can be argued that local environments and industrial networks are essential to an understanding of the spatial patterns and rhythms of technological innovation. To adopt an environment-based approach, as is the choice in this book, therefore involves considering each type of local or regional environment as a model of technological penetration. Why do some environments innovate more than others? Why do innovative environments sometimes cease to innovate?

The hypothesis adopted here is, then, that local environments play a major if not determinant role as incubators of innovative activity, as prisms through which stimuli to innovation must pass, as networks of interactions channelling and shaping the imprint of technological change in particular areas. The firm is not an isolated agent of innovation: it is one element within the local industrial milieu which supports it. The historical evolution and characteristics of particular areas, their social and economic organisation, their collective behaviour, the degree of consensus or conflict which characterizes local society and economy, these are the major components of innovative behaviour. Taking the local environment as the starting point for the analysis of the spatial impact of new technologies focusses attention on local industrial networks and linkages, on the internal capacity of the environment to generate industrial development.

During the 1950s and 1960s, a decline was observed in the intensity of local industrial linkages in many European countries which seems to have reflected new forms of industrial organisation. The growth of large firms was accompanied by increasingly internalized systems of inter-industrial relations, and a declining demand for "external" supporting activities. Falling transport costs allowed such firms to transfer semi-finished products and components from one factory to another over considerable distances (Keeble, 1976, 187-91). In most countries, the role of small and medium-sized firms as external suppliers and inter-linked producers inevitably declined.

The emergence since the early 1970s of new industrial regions based on high-technology activities has generated renewed interest in the problem of spatial industrial concentration. What is the underlying logic or justification of these new concentrations? Are these new development poles based on new systems of organisation of large corporations? Do they symbolize a resurgence in small and medium-sized firms and of linked industrial complexes? Does spatial proximity afford the same advantages today as it did in the past?

Several sets of reasons can be put forward to account for these new types of territorial industrial concentration, the variety of which may simultaneously reflect a single underlying logic, or stem from quite different basic causes. These are:

i) minimization of transport and distance costs, which may lead to a choice of nearby suppliers. Recent debate over the spatial implications of the growing significance of Japanese "just-in-time" production systems in Europe focusses on one special case of this motive for concentration (Sayer, 1985: see also Camagni's chapter in this book).

ii) constraints linked to the small size of the firm, which forces it to buy in from other firms numerous goods and services which it cannot produce itself at a competitive cost.

iii) clustering simply as a consequence of the capacity of certain local environments to generate large and growing numbers of new firms.

iv) access by high-technology firms to spatial concentrations of essential but scarce, mobile and environmentally-sensitive, highly-qualified research workers, scientists and engineers.

v) multiplier effects linked to the presence of a dominant successful company whose activity spawns profit-making opportunities for local entrepreneurs.

Underlying these processes are significant differences in approach between analyses focussed on costs and prices of production and those couched in terms of dominance and power relationships, as well as between static analysis of the contemporary functioning of territorial complexes and dynamic analyses of the formation and evolution of new industrial spaces over time. The question of the best approach to understanding contemporary trends cannot be resolved without recognition of the problem caused by the speed of current technological change, and hence the vital need for adaptation, for flexibility of production and organisation, for access to know-how and information. In fact, two opposing viewpoints exist which need to be synthesized: the approach which focusses on firm relationships, and that focussed on the local environment.

This is not simply a question of contrasting the multi-spatial logic of the large corporation with that of the small firm, more rooted in its original environment. Certainly, the spatial division of labour approach appears to offer a very convincing argument in claiming that the financial capacity of large firms, their ability to manage activities efficiently over considerable distances, and their control of markets and technologies, engender a form of territorial behaviour which rings the death knell of traditional industrial complexes organised around networks of linkages. The large firm is able to split its activities up into homogeneous functional units, and to restructure national - and international - space along centre-periphery lines, differentiated into core areas of research and decision-making and peripheral areas confined to production functions, routine activities and less-skilled work. In contrast, the small enterprise, firmly embedded in the local milieu which produced it, usually depends on this milieu for its technology, its markets, its labourforce. It develops most characteristically in those local environments which offer a mix of local activities and complementary services, and rarely engages in long-distance interactions.

These two approaches appear however somewhat simplistic alternatives when set against recent empirical observation. How for example do they square with findings of industrial complex dynamism based on new technologies and new firms, and the rise of some of these complexes and firms to national and global importance? What of recent assertions that inter-firm linkages are of fundamental and increasing, not diminishing, importance for competitive success in high-technology industry in the later 1980s (Gordon, 1987)? And what of the problems identified with traditional product-cycle spatial dispersion notions in such important examples as that of the US semiconductor industry (Castells, 1984, 1985: Saxenian, 1983: Glasmeier, 1985a: Gordon and Kimball, 1986: Scott, 1987: Scott and Angel, 1987)? In the latter industry the post-1970 period has witnessed processes both of increasing product maturity, accompanied by the classic phenomena of standardisation, the growth of large firms, and spatial dispersion, and of rapid technological change, accompanied by continuing new firm formation and territorial concentration. Silicon Valley, for example, now contains 25% of all US semiconductor employment, compared with only 10% in the 1960s. In other words, it seems clear that in a highly-complex and changing world, the logic of large-scale production, standardization and automation is only one of at least two possible strategies. An alternative, which seems to be developing in parallel, is customized production of integrated circuits for specific clients, involving a search for specialized market niches, and vertical disintegration. While steeply-rising costs of research and new model development undoubtedly favour giant firms in many high-technology sectors (Gillespie, Howells, Williams and Thwaites, 1987), small and medium-sized firms are also achieving success and competitive efficiency through specialization and customization (Keeble and Kelly, 1986: Gordon, 1987).

Thus for smaller high-technology firms, very close market contact and technological expertise in a specific and important product niche can provide the basis for survival and growth. Such a policy however automatically implies a need for linkages with, and ideally close proximity to, numerous associated firms. The two alternative strategies outlined above seem to have developed simultaneously, depending on the level of national and international competition, on the growth of market demand, on the respective size of markets for standardized and customized products, and on the ability of subsequent imitators to master large-scale production of standard products. This is well illustrated by the semiconductor case. Japan, as well as now increasingly independent producers in the Newly-Industrializing Countries, have caught up with even the most technologically advanced (VLSI: very large scale integration) US producers of standard components. In an intensely-competitive market-place, future survival demands complex corporate strategies. One approach may involve acquisition and integration into a large corporation of formerly independent producers, as with IBM's partial absorption of Intel and its development as a producer of specialized components. Another may be increased institutional control of lower-order markets, which allows some escape from the iron law of competition in the semiconductor industry: Texas Instruments for example now uses its own components in the electronic products it manufactures.

If such strategies are not feasible, survival necessitates the establishment of co-operative links with other firms in order to create an integrated collective capacity for innovation and control of technological development. It is at this point that spatial and industrial logic combine. As noted earlier, the 1960s semiconductor industry trend of territorially-concentrated but industrially-dispersed small firm development focussed on Silicon Valley gave way in the 1970s to a contrasting process of large-firm industrial concentration but territorial dispersion. By the 1980s, however, the evolution of production facilities, and notably assembly plants, set up earlier in peripheral areas such as Scotland and South-East Asia had reached a point where integrated development had become possible. Scotland has thus begun to develop its own innovative capacity (Haug, 1986), including some degree of indigenous new firm spin-offs (Keeble and Kelly, 1986), Singapore and Hong Kong are no longer simply centres for assembly and re-export operations, and even Manila and the Philippines now possess the beginnings of an electronics industry complex (Henderson and Scott, 1987). Similarly, the Eurotechnique plant at Rousset near Aix-en-Provence, which is simply a production facility, has nonetheless stimulated the creation of other micro-electronics firms in the vicinity and the beginnings of a complex. Thus the apparently opposing logic of large firm "de-territorialization" and SME local milieu development interact and overlap in subtle ways with each other, neither being automatically or invariably the more dominant.

The local environment, then, provides a vital basis or foundation for technological innovation in the 1980s. However, should it be viewed as a passive supplier of essential location factors for high-technology firms; or is it more appropriately treated as an active agent, provoking and stimulating new firm creation and existing firm modernisation by virtue of its inherited socio-economic structures and processes?

The latter viewpoint perhaps accords best with the progressive evolution of increasingly complex territorial structures. Within the broad global framework set by the process of large firm capital concentration, vertical integration and territorial dispersion, which produces the spatial division of labour, processes of spatial concentration of high-technology industry focussed on local environments can readily be observed at both ends of the chain. At the "centre", processes of specialization and customization of production based on co-operation between small and medium-sized firms reinforce local networks, restructure existing zones such as Silicon Valley, and create new concentrations such as Santa Cruz county in California (Gordon, 1987) and the Cambridge region in the UK (Keeble, 1987b). In the "periphery", technology-based complexes are beginning to emerge in certain cases such as Scotland and the Midi of France from the impact of large-scale production activities. In both situations, large and small firms are evolving sophisticated relationships, as for example through the sharing of the market between producers of standardized, mass-produced commodities and those of customized, small-batch products, or through the role of SMEs in investigating new approaches or products of potential value to larger local firms. The changing nature of sub-contracting also illustrates this well. Traditional sub-contracting links usually involved close control by the client firm based on precise instructions and specifications to its subordinate. Increasingly, however, small and medium-sized high-technology firms are operating with greater independence, providing large companies with both simple and complex components reflecting competitive advantage through specialization and particular technological expertise.

Of the many questions posed by current trends, one of the most interesting concerns the development of small and medium-sized firms. The substantial resurgence in numbers of such firms since the early 1970s in virtually all west European countries because of increased rates of new business formation has been charted by a number of studies (Keeble and Wever, 1986). However, considerable caution needs to be exercised in interpreting the significance of this trend, in spatial and wider terms. The small firm is by no means always the same thing as local decision-making power. While new firm creation does appear powerfully to be influenced by the nature of the local environment and the encouragement to entrepreneurship which that may provide (Keeble and Wever, 1986), the ability of small or medium-sized firms to maintain their independence is more problematic. Large firms rarely try or need to fight small firms: instead, they make use of them for their own benefit, in a great variety of ways (Aydalot, 1987). By such arrangements, including of course direct acquisition and takeover, they are able to combine scale advantages such as financial resources and organisational power with the flexibility and adaptability of the small company. The extent of local autonomy even in environments characterised by large numbers of high-technology SMEs is therefore a matter of debate.

Finally, what is the significance of geographical proximity with regard to local industrial development in the Europe of the 1980s? Is the growth of new high-technology concentrations, of innovative local environments, the result of a search for proximity? Of course, it is clear that proximity can no longer be reduced as in neo-classical economics simply to the need to minimize transport costs. The time when smallness of firm size linked with high unit costs of transport prevented industrial dispersion is long since past. Large firms, with their ability to organise and manage production facilities over large distances, have clearly benefitted from decentralization rather than been inhibited by increased transport costs.

The significance of proximity lies rather in the context of the efficient functioning of certain activities essential to competitive success in advanced-technology industry, and of certain systems of inter-firm relationships. The most obvious example of the former is research and development activity, with its need for frequent research contacts, for intellectual ambience, and for personal relationships, and therefore for proximity to other researchers. In the case of the latter, co-operation and specialized linkages between small and medium-sized firms often impose on them very rigid relationships with customers, suppliers and sub-contractors. Geographical proximity is therefore only the visible expression of an underlying functional network. The two approaches of industrial or firm structure on the one hand, and local environments on the other, cannot therefore be regarded as based on a different logic. To a considerable extent, the local environment is the material expression on the ground of inter-firm functional relationships.

THE GREMI AND EUROPEAN REGIONAL DEVELOPMENT

The research carried out in the last few years by the GREMI (Groupe de Recherche Européen sur les Milieux Innovateurs) enables a general picture to be constructed of current innovation behaviour and technological change in the regions of Europe. In this respect, as in many others, the continent's old industrial regions in Britain, Belgium or northern France contrast with newly-industrialising regions such as the French Midi or East Anglia. Different economic and social systems define very different rules of the game in different areas; Swiss decentralisation contrasts with French centralisation, western European market economies contrast with eastern European planned economies. Particular regions suffer from the specific problems, but also benefit from certain advantages, of a frontier situation. In one region, the industrial base depends upon heavy industry and a small number of giant firms, while in another, complex industrial structures link high-technology leader firms with numerous smaller businesses.

In the course of its work, the GREMI has investigated the impact of innovation and new technologies on industrial change in some 15 European regions (Aydalot, 1986). These include old heavy industrial regions (Newcastle, Charleroi, Poznan, St Etienne), traditional industrial regions based on lighter, small-firm industries (watch-making and mechanical engineering in the Swiss Jura and Besancon), less-industrialised but developing regions (Aix-en-Provence, Côte-d'Azur, the Ticino region of Switzerland), diversified metropolitan regions (Paris, Milan, Amsterdam), and peripheral semi-industrial regions (northern Greece).

The ease with which these regions are adapting to or embracing a development model based on new technologies and new activities varies in each case. In regions such as Newcastle, longstanding domination by externally-controlled corporations severely constrains local industrial restructuring: in Charleroi, in contrast, the arrival of a branch unit of a large externally-based company introduces new technologies and innovative products, and creates new market opportunities for local firms in a declining economy. The St Etienne case illustrates emergence in some regions of a collective development initiative, involving different local actors and organisations, both public and private, and the creation of new networks. The progress of such initiatives is however often slow and of long-term rather than short-term significance, especially where an all-powerful central state insists on limiting local decision-making powers.

Europe's leading metropolitan regions are invariably better placed than the heavy industrial regions exemplified above. They represent major concentrations of initiative, enterprise and financial resources. They possess a powerful base of scientific and industrial research and knowledge. Their role as the major national centres of decision-making and of market demand confers on their industries and firms significant advantages of adaptability and flexibility, including technological flexibility (De Jong, 1987). These advantages are further enhanced by the diversity of their industrial activities, the presence of technologically-advanced sectors, and the absence of domination by only one or a few giant firms, even in such cases as Milan and Turin with their leading companies of Pirelli and Fiat.

Those older industrial regions where sectoral specialization did not involve concentration into a few large firms face a specific but different set of problems. Massive and painful restructuring is often necessary: but at the same time, they do seem to possess an innate flexibility which permits substantial technological change. Though resistance to such change exists, these regions have demonstrated a capacity for adaptation which can take unexpected forms. Thus some firms have kept their traditional technology but switched production to a new sector, whereas others have adopted new technologies but continued manufacturing within their traditional sector. Whether or not existing organisational structures need to be modified to assimilate new technologies represents another area of uncertainty. Quite often it is a subtle mix of old and new which provides the key to successful integration of radically new technologies. Entrepreneurial and management skills, and flexible labour force attitudes, are crucial in achieving this optimal mix.

Paradoxically, the most revolutionary processes of technological change appear to operate best in previously least-industrialised areas. What industrial activity was there only 40 years ago in California's Santa Clara or Orange counties? Of what industrial significance were Cambridge, Antibes or Aix-en-Provence in the 1950s? Of course, only a relatively few, favoured localities have experienced rapid technology-based industrial growth, out of many less-industrialized regions. So what particular combination of factors, of existing knowledge, research capacity, "quality of life", or entrepreneurial or corporate initiative, has enabled these and other such areas suddenly to embrace the fifth Kondratieff industrial revolution, having been bypassed by the first four?

It is this diverse pattern of European regional industrial and technological change which provides the compelling focus of the GREMI's work, key insights and themes from which are presented in the following chapters.

REFERENCES

Aydalot, P. (1976) Dynamique spatiale et développement inégal, Economica, Paris

Aydalot, P. (1983) La division spatiale du travial. In J. Paelinck and A. Sallez (eds), Espace et localization, Economica, Paris

Aydalot, P. (ed.) (1986) Milieux innovateurs en Europe, GREMI, Paris

Aydalot, P. (1987) The role of small and medium size enterprises in regional development: conclusions drawn from recent surveys. In M. Giaoutzi and P. Nijkamp (eds), Small and Medium Size Firms and Regional Development, Croom Helm, London

Castells, M. (1984) Towards the informational city? High technology, economic change and spatial structure, Institute of Urban and Regional Development, University of California, Berkeley, Working Paper 430

Castells, M. (ed.) (1985) High technology, space and society, Sage, Beverly Hills

Czamanski, S. (1971) Some empirical evidence of the strength of linkages between groups of related industries in urban-regional complexes, Papers of the

Regional Science Association, 27, pp. 137-60
De Jong, M.W. (1987) New economic activities and regional dynamics. Unpub. Ph.D. Thesis, University of Amsterdam: see especially chapters 5 and 6, New firms and high technology in the Netherlands, and New firms in Amsterdam
Freeman, C. (1986) The role of technical change in national economic development. In A. Amin and J. B. Goddard (eds), Technological change, industrial restructuring and regional development, Allen and Unwin, London, pp. 100-14
Freeman, C. and Soete, L. (1985) Information technology and employment, Science Policy Research Unit, Brighton
Gillespie, A., Howells, J., Williams, H. and Thwaites, A. (1987) Competition, internationalization and the regions: the example of the information technology production industries in Europe. In M. J. Breheny and R.W. McQuaid (eds), The development of high-technology industries: an international survey, Croom Helm, London, pp. 113-42
Glasmeier, A. (1985a) Spatial differentiation of high technology industries: implications for planning. Unpub. Ph.D. dissertation, University of California, Berkeley
Glasmeier, A. (1985b) Innovative manufacturing industries: spatial incidence in the United States. In M. Castells (ed.), High technology, space and society, Sage, Beverly Hills, pp. 55-79
Glasmeier, A., Hall, P. and Markusen, A. (1983) Recent evidence on the spatial tendencies of high technology industries: a preliminary investigation, Institute of Urban and Regional Development, University of California, Berkeley, Working Paper 417
Gordon, R. (1987) Growth and relations of production in high-technology industry. In M. Breheny and P. Hall (eds), The growth and location of high-technology industries: Anglo-American perspectives, Rowman and Littlefield, Totowa, New Jersey
Gordon, R. and Kimball, L.M. (1986) Industrial structure and the changing global dynamics of location in high technology industry, Silicon Valley Research Group, University of California, Santa Cruz, Working Paper 3
Hall, P. (1981) The geography of the fifth Kondratieff cycle, New Society, March 26, pp. 535-7
Hall, P. (1984) The geography of the fifth Kondratieff. In P.

Hall and A. Markusen (eds), Silicon landscapes, Allen and Unwin, Boston, pp. 1-19
Haug, P. (1986) US high technology multinationals and Silicon Glen, Regional Studies, 20, 2, pp. 103-16
Henderson, J. and Scott, A.J. (1987) The growth and internationalisation of the American semiconductor industry: labour processes and the changing spatial organisation of production. In M.J. Breheny and R.W. McQuaid (eds), The development of high-technology industries: an international survey, Croom Helm, London, pp. 37-79
Hymer, S. (1979) The multinational corporation and the international division of labour. In S. Hymer, The multinational corporation: a radical approach, Cambridge University Press, Cambridge, pp. 140-64
Keeble, D. (1976) Industrial location and planning in the United Kingdom, Methuen, London
Keeble, D. (1987a) Industrial change in the United Kingdom. In W.F. Lever (ed.), Industrial change in the United Kingdom, Longman, Harlow, pp. 1-20
Keeble, D. (1988) High-technology industry and local economic development: the case of the Cambridge phenomenon, Environment and Planning C, Government and Policy, forthcoming
Keeble, D. and Kelly, T. (1986) New firms and high-technology industry in the United Kingdom: the case of computer electronics. In D. Keeble and E. Wever (eds), New firms and regional development in Europe, Croom Helm, London, pp. 75-104
Keeble, D., Owens, P.L. and Thompson, C. (1983) The urban-rural manufacturing shift in the European Community, Urban Studies, 20, 4, pp. 405-18
Keeble, D. and Wever, E. (eds) (1986) New firms and regional development in Europe, Croom Helm, London
Markusen, A. (1985) Profit cycles, oligopoly and regional development, MIT Press, Cambridge, Massachusetts
Massey, D. (1979) In what sense a regional problem? Regional Studies, 13, 2, pp. 233-44
Perroux, F. (1955) Note sur la notion de "pôle de croissance", Economie Appliquée, 8, pp. 307-20
Rothwell, R. and Zegveld, W. (1982) Innovation and the small and medium sized firm, Frances Pinter, London
Saxenian, A. (1983) The genesis of Silicon Valley. In P. Hall and A. Markusen (eds), Silicon landscapes, Allen and Unwin, Boston, pp. 20-34

Sayer, A. (1985) New developments in manufacturing and their spatial implications, University of Sussex, Urban and Regional Studies Working Paper 49

Scott, A.J. (1987) The semiconductor industry in south east Asia: organisation, location and the international division of labour, Regional Studies, 21, 2, pp. 143-59

Scott, A. and Angel, D.P. (1987) The US semiconductor industry: a locational analysis, Environment and Planning A, 19, 7, pp. 875-912

Swyngedouw, E. A. and Anderson, S. D. (1986) The regional pattern and dynamics of high-technology production in France. In Association de Science Régionale de Langue Francaise, Technologies nouvelles et développement régional, Centre Economie-Espace-Environnement, University of Paris 1, pp. 441-59

Thompson, C. (1987) High technology development and recession: the local experience in the US, 1980-2, Environment and Planning C, Government and Policy, forthcoming

Vernon, R. (1966) International investment and international trade in the product cycle, Quarterly Journal of Economics, 80, pp. 190-207

Chapter 2

Technological Trajectories and Regional Innovation in Europe

Philippe Aydalot

INTRODUCTION

The introduction of new technologies in the 1960s and 1970s focussed attention on certain privileged areas such as Boston's Route 128 or Silicon Valley in California. While traditional industrial areas were able to achieve some degree of technological progress, the major trend visible until recently seemed to be the emergence of areas characterized by an interaction between science and industry leading to a massive industrial development which created a huge concentration of new activities in virgin areas. The considerable prestige of the examples cited has however been overshadowing other innovatory events to the extent that the capacity for innovation of particular areas is often assessed merely in terms of the processes observed in Santa Clara county. The existence of university research capacity, of firms originating through university efforts, of venture capital development, or of local entrepreneurial spirit, are all too easily adopted as universal indicators of technological development, in the M_4 corridor as well as in Sophia Antipolis or in Cambridge. However, this is surely too restricted an approach, given that innovation is known also to follow other patterns of behaviour. A region's ability to innovate and develop new technology is conditioned by many factors, not just whether or not its structure is similar to that observed at Palo Alto! Many innovation patterns and paths exist; there is no single route to regional technological development. It is important to identify these different trajectories and to find out what are the elements they need in order to develop. An attempt will be made in this chapter to prepare the ground for this issue by drawing on some of the results of research carried out by GREMI (Aydalot,

1986), which provides a detailed analysis of some 15 regions in Europe.

Three types or patterns of innovation can be distinguished, and every local environment or milieu should know how to adapt to one of them so as to innovate. The first is the restructuring of a pre-existing industrial environment: local firms facing the risk of a serious decline are able to renew and regenerate their activities through the agency of technological change. The second type is large firm corporate restructuring. In this case, large companies are forced to introduce new products and adopt new processes in order to maintain their market share and monopoly power. As a consequence, the firm may modify its location system, concentrating some of its activities and dispersing others. The third type involves the "production" of knowledge and its direct application into manufacturing by individual entrepreneurs coming from a research background and setting up their own firms. In this case, the change is a more fundamental one: the firms which apply it are, for the great majority, small and recently formed companies while the organization of the industrial area is totally different from that observed in trajectories mentioned above.

According to each of these patterns, different actors, different innovations and different processes are at play. Assets needed in order to innovate according to one model would be superfluous in another one. Not every area is predisposed to offer the terrain for any kind of innovation. How to identify the processes through which each of the three patterns are embodied, how to determine for each area the technological path to which its structure is best suited, are the questions to which this chapter will attempt to provide some answers.

After a few comments on technological innovation, the chapter will consider the major mechanisms which nowadays differentiate local level innovation. The processes structuring major technological trajectories will then be examined, before presentation of a typology of actual innovation areas. The chapter will conclude by demonstrating the importance of a dynamic analysis of innovation areas.

SOME ASPECTS OF TECHNOLOGICAL INNOVATION

Whatever the sector in which innovation occurs, whatever

the type of commodity to which it is applicable, or whichever form it takes (i.e. product innovation or process innovation), the fact remains that innovation entails a major upheaval affecting all aspects of the running of a firm. To modify technology is equivalent to modifying production organization as well as the equipment and the hierarchies present within the firm. It means that new workers possessing the required skills will be employed, new forms of management will be adopted and new markets have to be sought, at least to a certain extent.

Technological innovation therefore implies a chain of changes in order to achieve a new kind of coherence. This in turn may require complicated adjustments which can turn out to be all the more difficult if the enterprise is a large and an old one. If the technology involved is revolutionary, the resistance to change in a large firm will be strong. Managers may thus be tempted to marginalize the importance of innovation in their production activity, preferring to fall back on tried and tested market and product niches. Even if the firm has the basic knowledge needed in order to adopt an innovation, it will often keep it in stock, so to speak, until an immediate need for it arises (Castells, 1984, Aydalot: 1986). Certainly, the large enterprise does innovate continuously: but it cannot afford radically to upset all of its activities at the same time. It has to plan its activities over long and medium-term perspectives so as to guarantee its survival, to manage its labour force smoothly in the context of often-strict labour agreements and regulations, and to take into account different constraints which operate on a global scale. All these considerations prevent the large firm from making overnight changes simultaneously in products, processes, markets and the labour force.

In contrast, small and medium enterprises (SMEs) can, particularly in their early years of life, show a more marked tendency for radical innovation (Rothwell and Zegveld, 1982). In general terms, it is probably true that old firms innovate less than new ones, and large firms innovate in a more regular and steady way than small ones. This calls for some original relationships between small and large firms during an intense innovation process. Thus, for example, small enterprises may play a vital innovatory role in the context of local economic crisis. When a technological challenge from outside begins to ruin the local industrial milieu, a large firm can turn out to be too rigid to integrate a new technology quickly enough and consequently is forced to close. However, this may in turn give rise to new firms which are set up by its former workers. This was how the Besançon firm of Lip, which had irreversibly declined, was reborn in the form of half-a-dozen new local enterprises set up by redundant engineers from Lip (Pottier and Touati, 1986). The new firms managed to adapt to the new market conditions.

However, the upheaval is never complete so innovation always comes as a combination of some already familiar forms of organisation. Thus Silicon Valley is an example of an industrial complex which is classic in its structure: its firms are of modest size (at least in the early stages of their development), are very homogeneous in sectoral terms and are characterized by a direct social relations system avoiding unionization. In this case, an apparently very traditional form of collective organisation, which has been viewed as outdated for the last 40 years, lives side by side with some very advanced technologies. The innovative firm adjusts its functions so as to create a coherent system which is often embodied in traditional forms (sub-contracting for example). The latter are however used in a new way and in original combinations.

Effective technological innovation thus demands a total adaptation of the components in a local economy system. Firms, entrepreneurial attitudes, social relations, inter-firm relations should all act in unison with requirements dictated by the spread of new technology in firms and local environments. This will allow advanced technology to penetrate into areas which possess the capability of making the right adjustments. An inflexible area cannot incorporate new technology even if, technically speaking, nothing prevents it from doing so. A new approach to understanding innovative environments can thus be put forward based on their former structure. What will be their capacity for change? Which modern organization pattern will they be able to adopt?

SPATIAL PROCESSES OF INNOVATION DIFFERENTIATION

In this section, stress will be laid on three aspects of local environments which may significantly influence the adoption or creation of technological innovations.

The nature of local relations

The historical development of manufacturing in Europe over the past century has often involved the creation of industrial complexes through macro-unities (growth poles), sometimes in a less structured fashion. This process included the establishment of inter-industrial relations, such as the exchange of intermediate goods and semi-products, and the evolution of local industrial milieus or environments characterised by more or less integrated networks of inter-firm linkages. The most typical situation was sub-contracting whereby inter-industrial relations assumed a quasi-institutional form uniting, in an uneven relationship, dominant firms and those small sub-contracting firms which develop in their shadow. Since the appearance of new science-based complexes, however, a different kind of relationship has been emphasized. Here, technical change has opened up a personal relations channel, namely contacts between firms and universities and large private and public research centres, relationships with technical centres of collective research, inter-firm technical co-operation and, above all, direct relationships between engineers and scientists through informal contacts, especially among engineers laid off in one enterprise and re-employed in another one (Planque, 1985).

A third type of local relations, of a more general nature, should not be omitted from this discussion. This is the social consciousness present within an area, the creation of a concensus which may help spur on the progress of an entire region. In many European regions at the present time, milieus are being formed, stimulated by the local Chamber of Commerce or other organisations, associations are being set up uniting the heads of local firms, and local networks are being created by local agencies striving to promote technology. Even universities sometimes build up contact structures through which their influence, research and technology can be spread throughout an industrial area.

A contrast can thus be drawn between the classic pattern of inter-industrial relations and sub-contracting, and the modern pattern based on direct, inter-personal relations. This, however, would be to oversimplify the contrast. Indeed, recent investigations have shown that inter-industrial relations remain important. A detailed analysis of innovation processes among firms of the Paris area (Decoster and Tabariès, 1986) revealed that establishments in the southern suburb constitute not so much a "technopolis" (which means a set of industries oriented towards science and technology), as an industrial complex often run in a traditional manner which, nevertheless, has a specialisation in advanced technology sectors. This is to say that the network of relationships that feeds and structures the area mainly involves inter-industrial relations, in the form of exchanges of goods which provide the stepping stones to innovation (the dominant firm imposes a new pattern or a new process, the purchaser creates a demand for a new product, the firm provides a product which then enables the operation of a new process).

While certain regions have been able to innovate giving priority to knowledge transmission based on an exchange of brain-power or a direct exchange of process technology, it should be remembered that there is also the classic form of exchanging technical know-how as a direct product of normal inter-enterprise market relations. The industrial complex is, undoubtedly, the most common form of an industrial milieu. It has been shown (Gordon and Kimball, 1986: Scott and Angel, 1987) that the most advanced technological complexes in the U.S.A. (Orange county, Silicon Valley, Santa Cruz county) developed from renewed forms of sub-contracting. Small firms which exploit very specialised high-technology niches are often forced to attain a high degree of specialisation, and to sub-contract a large proportion of their production in order to save the trouble of costly research in a field that is not within the scope of their competence. Moreover, it is commonplace for large dominant enterprises to avoid technological effort in what may be an insignificant field for them, and to limit their domination to placing orders for sets of composite products, rather than for components whose technical specifications they have already determined themselves. Thus, sub-contractor firms should not only attempt to master the technology of their products without having to consult their dominant purchasers, but they should moreover associate with other similar companies to supply entire, assembled, components as is the case, for example, with the automobile industry in the Paris area.

The operation in new ways of such classic processes as sub-contracting helps to maintain the coherence of local industrial complexes and networks during periods of commercial and technological uncertainty and rapid change. Indeed, in spite of apparent shifts, it is doubtful whether industrial structures have actually changed a great deal in recent years, for large firms remain at the centre of the industrial stage, perhaps to an even greater degree than ever before. However, they do seek to exploit as much as possible the advantages of flexibility which arise from their multi-spatial character. Thus, the only areas which will appear to be innovative will be those which prove to be capable of adapting the ways in which local industrial networks operate to this constraint.

Labour markets

A rigid labour market which was established a long time ago, is based on historic rather than evolving skills, and is run according to a codified social relations system, is often regarded as an obstacle to changes required by technological innovation. In contrast, new, not fully established, labour markets are likely to be much more flexible and thus more readily adaptable to new needs. Old industrial environments generally have "closed" labour markets operating under strict rules resulting from unified working procedures and standardized guarantees negotiated with labour unions. In difficult periods firms and organizations in such environments are not likely to evolve which results in defensive labourforce behaviour. This type of closed labour market is mainly found in areas of large-firm domination. However, when the industrial system is based on small enterprises, relations remain more personalized and guarantees given to workers are limited due to the firm's own intrinsic vulnerability. During periods of change it is common for many enterprises to disappear which, in turn, liberates, as it were, the labour market from the "rules of the game" which prevailed before. A fortiori, in areas of no industrial tradition, firms create their own labour markets and are merely subject to national legislation.

The research carried out by GREMI on regional industrial labour markets in Europe reveals three patterns of development:

i) A labour market which is formed in a "virgin" area, whereby the workers migrate in large numbers to the area from elsewhere. As a result, they do not repeat or imitate their parents' jobs, qualifications and rules of the social game. They do something else, elsewhere and in a different manner. A good example is the Sophia Antipolis labour market in the Côte d'Azur.

ii) A labour market which has been broken up and then partly re-utilized by new employers. In the Swiss Jura, for example, almost two-thirds of the jobs in watch-making disappeared in only 15 years as well as 60% of the enterprises. Redundant workers often left the area and their occupations to take up a new job in the tertiary sector in Geneva, for example. This kind of restructuring in turn creates new possibilities for new firms, which result from local entrepreneurship and inward investment, to set up a new labour market with different qualifications, different techniques, and new rules of the game. Thus a labour market has been created which only partially re-uses former know-how.

iii) When the industrial system finds itself unable to adjust to deepening economic crisis and hence falls into decline, the labour market on which it depended also undergoes a slow but sure decline. Eventually, it is likely to be destroyed without any recycling of expertise or traditional skills. Next to it, a new labour market will emerge which is separate from the former one; it is merely located alongside it and is formed by employing workers from outside the previous industrial system. Firms from outside the region arrive with their own rules and techniques; they survive, grow and recruit new workers, deliberately avoiding any direct connection between the previous labour market and the one to which they are now contributing (Newcastle, Charleroi).

Development through supply or demand

On a different level of analysis, an interesting distinction can be drawn between development stimulated by aspects of the supply capacity of a region and by the market. Recent experience has led economists to emphasise the importance of regional capacity characteristics for the emergence of a set of firms which brings with it such assets as knowledge, industrial initiative and technological innovation. In the Silicon Valley type, the process starts with knowledge. In this example, firms were indeed sometimes set up to take advantage of research-based innovations even though there was no immediate demand for their products; thus, supply precedes demand, the ability to produce comes before production. In the case of old industrial regions which are undergoing restructuring, the same approach seems to be appropriate, in that the capacity of firms to be restructured and to innovate seems to be essential for a successful reconversion.

However, it should be borne in mind that sometimes it is demand which will reshape, if not create, an industrial milieu. Two examples illustrate this clearly. Defence spending and the development of public research can act as a triggering mechanism for technological development as was the case in California. Demand is, therefore, present before innovation and it is the former which is indeed responsible for the creation of the industrial milieu. Even if, in the famous example of Silicon Valley, a very particular structure has enabled rapid industrial development to take place over the last 30 years, it cannot be denied that the demand engendered by massive and specific public expenditure has played the major role. As Gordon and Kimball (1986) point out, "Far from being created by ineffable scientific genius and heroic entrepreneurialism, the electronic industry in its infancy proceeded ultimately from extensive State intervention". In the 1950s, military demand was both considerable and very selective in choosing the most advanced technology; price was only of secondary concern. This was the vital stimulus enabling the establishment of the technological superiority of California, and particularly Silicon Valley. Moreover, this superiority was in turn the basis for industrial development during the 1960s and 1970s. Though it may seem obvious that the region in question had for the most part the great advantage of early research leadership, the essential role of new demand and new markets must not be overlooked.

Another example also illustrates clearly the importance of initial demand. The concentration of high-technology activities in the southern suburbs of Paris is, to a large extent, related to the fact that there exists a strong demand for scientific instruments from universities, scientific colleges (French "grandes écoles") and research centres in the area. It was this demand which, in due course, created the corresponding supply which was established close at hand. Similarly, the importance of sub-contracting as a privileged mechanism for the development of advanced-technology firms confirms the importance of the impetus that demand gives preceding innovation. Interestingly, the distinction between development from knowledge or from demand is also related to some degree to differences in the behaviour of small and large firms. The small, innovative firm is able to precede demand to exploit a possibility opened up by scientific research whereas the large one innovates to be better suited to supply a pre-existing or predicted demand. Here again this generalization must be qualified; no firm can be created, no new item introduced without a specific demand for it or without such demand being clearly anticipated by the entrepreneur. The split between innovative firms and others is, therefore, most probably based on their capacity to distinguish demand when it is not yet clearly formulated or when it has only appeared very recently. A technologically innovative area would, therefore, be one which enables the spread of new demand, indeed which creates demand while at the same time informing the potential demander about the capacity of supply of local innovative firms. An area like Silicon Valley is therefore innovative not merely because of its capacity for new supply, but also because of its concentration of new demand.

Environmental interaction and innovative enterprises

The innovation process activates all the processes described so far. Not all are necessary for innovation, and every technological path can be defined through those relationships and processes to which it gives priority. These relationships, however, reveal the collective character of innovation. Table 2.1 illustrates the role of the environment in influencing a firm which innovates, the environment acting as a supplier of inputs, of knowledge, of an innovatory atmosphere, and as a market for goods. Most of these relationships can however be internalised by the firm if it is large enough and if there is interest in doing so. Vertical integration frees the firm from the need for contacts with other firms, while research development within the firm can reduce technical dependency on external agents or maintain an existing monopoly of technical knowledge. The cost of such a strategy may however be considerable.

TECHNOLOGY TRAJECTORIES

Let us recall the three major patterns of technological innovation suggested earlier:

i) The restructuring of an industrial milieu. Here the main agent is the engineer or skilled worker who acts within

Table 2.1. The Environment of the Innovative Firm

Type of relations	External to the firm			Internal to the firm
	Partners	Market relations	Non-market relations	
labour	labourers	formation of a labour market	role of unions	internal training
	engineers and scientists	re-employment of engineers	informal contacts personal relations	inter-establishment mobility
	State		adaptation of training centres	
inputs market	firms	relations between suppliers and clients specialized sub-contracting co-operation between SMEs	associations, clubs Chambers of Commerce	vertical integration
scientific and technical knowledge	universities government research laboratories	research contracts market for scientific instruments	spin-offs personal contacts	R and D laboratories of the firm
	firms	specialized sub-contract co-operation between SMEs		
specialized services	service enterprises	venture capital		specialized departments of the large firm
innovation impetus	State	public spending		
	public agencies	contracts, subsidies		
	firms	sub-contracting	clubs, informal meeting points	

the production process. An innovation emerges from within the firm, utilizing existing inputs and labour skills, but also modifying and restructuring them to meet the needs of the new product or process. This represents the typical type of innovation in traditional industrial regions.

ii) Innovation in a large firm mastering a new field. Aware of new and developing markets, the large firm accumulates the essential knowledge needed for technological innovation which it then applies in its own R and D laboratories. It then begins production according to its needs and to a medium-term plan. This kind of process is not "localised" so that the innovation will be implemented in areas supplying the appropriate location factors.

iii) An innovative enterprise which is a direct product of research. In this case, knowledge is accumulated outside the firm by scientists who acquired their competence, not through experience in industry or from their knowledge of the market, but through non-industrial research. This innovation pattern implies a very solid scientific basis and a technology which is so new as to rule out any advantage from past experience. Underlying these three patterns are three different spatial processes. The first innovation pattern (reconversion of industrial milieu) involves adjustments which take the form of a complex breaking-branching process. The second (planned innovation associated with large firms) implies that the areas in question are able to attract firms from outside. The third (establishment creation based on scientific research) results in a process of original polarization whereby the determining macro-unity is a knowledge centre and not an industrial enterprise.

The breaking-branching process

Is it possible for innovation to emerge from the very core of a traditional industrial milieu? Faced with the risk of a major crisis, old industrial regions often prove to be capable of adapting to the demands of new technology, of renewing their organisation and their structure, of overcoming a technology timelag if not actually taking the lead in new technology development. The success of the "swatch" clearly shows how an industrial area forged by an outdated technical and commercial tradition was, after a period of severe recession, able to reproduce a system of production adapted to new demands, in terms of technology, products, markets, firm structures, and labour skills. The Swiss watch-making industry has thus successfully adopted a new technology which was previously its rival. This was achieved by restructuring and re-employing its inputs and know-how, innovations coming from within local industry. From empirical investigation of numerous examples of this type of situation, it would appear that there is always present in the process of adaptation of old industrial milieus an element of breaking away from the past, and an element of branching out into new technologies and activities. This combination is, perhaps, necessary in order to succeed.

It must be stressed that the behaviour described above always represents a reaction to an external threat. As long as traditional organisational structures can be perpetuated, industrial milieus are confined to reproducing existing techniques and forms of organisation. It is perhaps only in reaction to the risk of complete ruin that technical (as well as organisational) innovations will occur. Thus Nijkamp's research in the Netherlands reveals that 33% of innovative firms studied had innovated because they were facing a collapse (Hoogteijling, Gunning and Nijkamp, 1985). In many cases, then, innovation is an enforced answer to a new, and externally-threatening, situation. When an organisational form is outdated, when firms and jobs disappear, local resistance diminishes so that innovation will be accepted even if previously valued assets are at stake. Indeed, innovation demands a great deal of breaking away, which can lead to its being opposed and prevented so long as such an attitude safeguards acquired assets. In some cases, even a serious industrial crisis does not break down barriers to change, and the local economy as a whole also declines. Recent dramatic employment losses in different European industrial regions provide a good illustration of just how powerful the processes of breaking away can be. In Newcastle, industrial employment dropped by 30% in the five years 1976 to 1981, with a 47% decline in textiles and a 32% loss in mechanical engineering and metallurgy; in Besançon, watch-making employment fell by 57% between 1975 and 1985; the industrial workforce declined by 6% per annum in the district of St Etienne in 1983 and 1984. In nine years, coal-mining in Charleroi lost 75% of its labour-force (1974-83). In Switzerland as a whole, the number of watch-making firms fell by over 60% between 1970 and 1985 (from 1620 to 600) while employment in this industry decreased by 65%.

Faced with crisis situations of this kind, modernisation

is likely to be accepted, even if it at first requires great sacrifices. Manpower in the watch-making industry of the Swiss Jura has never decreased so fast as between 1981 and 1984 (by 11% per annum as against 5% in the decade of the previous crisis): yet this was the very period in which the production of "swatches" began to prosper. The break-away process is accompanied by major disruption of the labour market. Moreover, not only does the traditional industrial labourforce disintegrate but, because of modernisation, new workers often take the place of old ones. In the French telecommunications industry, the labour force dropped sharply between 1977 and 1984 (from 75,000 to 50,000 workers), certain qualifications were lost while new ones were developed, and traditionally skilled jobs in assembling and cabling disappeared; thus, specialised jobs were hit. However, while technically-qualified employment involving traditional electrical engineering expertise declined by 30% over this period, demand for skilled electronic engineers increased rapidly, the proportion of such workers in the telecommunications industry rising from 9.6 to 16.9% of total employment. Finally, also from 1977 to 1984, the number of manual workers in the industry decreased by 50% (from 39,200 to 20,400) declining from 57% to 39% of total employment. This sector has thus undergone a major process of retraining and increasing productivity, coupled with a marked decrease in employment, developments which took place due to a major technological change, that is, the transition to the electronic telephone.

This very mobility of the labour market during periods of crisis enables necessary adjustments to occur. Firms are also affected by the break-away process. Thus over 1000 watch-making firms in the Swiss Jura were forced to close down in a period of 15 years, as were 25 of the 40 watch-making firms of over 10 employees in Besançon in ten years. Reconversion is therefore based on the ability of a minority of enterprises successfully to adopt a major technological change. Firms which survive because they were able to tackle a radical upheaval nevertheless often maintain some contact with what had existed before. Thus in the Jura watch-making region, market demand and experience provided the basis for diversification. Examples included a shift from mechanical to electronic skills, a gradual integration of electronic components - integrated circuits, diodes - adopted for watch-making into other products, combination of mechanical and electronic technologies as with the manufacture of robots, and the development of new products such as pacemakers which possess features and require skills similar to those in watch-making. When the founders of new firms engaged in these activities are considered, it is also clear that the great majority were already engaged in some professional activity in the Swiss Jura, and that almost all of them were already living there. Likewise, the Besançon experience shows that new firm spin-offs originate from existing firms (for example Lip) and are not of extra-industrial origin.

Hence, reconversion of industrial milieus rarely involves total and simultaneous upheaval in markets, products and technologies, all at the same time. Where there is an industrial milieu composed of small and medium-size enterprises employing workers with a high level of qualifications, and providing a certain technological continuity exists, an externally-imposed technological "aggression" can be absorbed and can indeed set off new development. The capacity of some of the firms to recycle existing local know-how into a new mix of the constituents of the enterprise (ie, its technology, its workers, its markets and products) determines the way in which the local area is able to innovate, and thus avoid or escape from the crisis.

Attraction

In the GREMI examples, regions such as Charleroi, Newcastle, St Etienne or le Valenciennois reveal how innovation can be delayed and the different forms modernisation takes.

In Newcastle and Charleroi in particular, a partial modernisation has been taking place. This development reflects the actions of large firms external to the regional environment. There is no recycling of local know-how nor are there any links with former specialisation. A new labour market is being formed with no connections with the previous one. A dual structure is gradually being built; the former structure is declining while a new one, quite separate from the old, is being superimposed upon it.

In these areas, all innovation of any importance seems to originate outside the area so that the local environment plays a minor role compared to external influences. The managers of the SMEs in the area often have limited engineering expertise, while their firms are very specialised and cannot be reconverted as it would require a significant change of technology. These local firms are used to living in the shadow of large, dominant companies which provide them with orders and techniques and whose decline is inevitably transmitted to the local supplier. Today, in the Valenciennois district (Viola, 1986), most innovative firms are subsidiaries of external firms. They depend on contacts with their customers, their suppliers and their parent company for innovating initiative, technical expertise and the form of innovation. The local environment is a negative factor. Unlike the traditional industrial milieus already discussed, which are characterized by the redeployment of local know-how, innovation here is a means of detachment from the local environment. The environment which previously had supported local enterprises is now no longer capable of doing so. Today, the only reaction of the local milieu is to generate awareness of the necessity for innovation because of an inherent fear of a growing crisis. It cannot however provide either techniques or markets, neither innovators nor the necessary labour. Qualifications cannot be recycled and attitudes to work need radically to be changed. Miners and steel workers were used to an organised work-style, requiring no flexibility or initiative, with a fixed working plan. This does not correspond to the mobile, changing needs of SMEs. The previous high salaries, security, and social advantages of workers in coal mines and iron and steel plants were assets which make it more difficult for these men to accept conditions offered by new SMEs. At all events, the type of skills required by innovating firms are not really available from the area's previously dominant activities.

If local assets are of little value, what is there to attract enterprises from outside the region? Factors which may play some part here include existing infrastructure, the generally good quality of manpower, locally-supportive attitudes towards new investment and enterprise, available space or existing industrial buildings, and even the location of the area, which may attract inward foreign investment aimed at serving the national market. In the latter case, as well as more generally, the role of government regional policy incentives may be considerable, with grants of up to 50% of total investment available in some regions. In the context of capitalist investment, such incentives may become a determining factor in location. In this type of situation, we are dealing with a particular type of innovation which results from attracting external firms, through the juxtaposition of a new developing milieu next to an old milieu which is declining without any attempt to innovate. It is dependent on large firms which are attracted by certain factors in the region, enabling them to operate production functions while retaining their decision and research centres elsewhere.

A quite different type of environment can also rely on imported technological developments, namely regions without an industrial tradition which offer attractive elements to large and sometimes small firms, their managers, entrepreneurs and workers. This type of situation is quite common. In some cases, the area of attraction is situated on the outskirts of a large industrial agglomeration (Orange county near Los Angeles, Berkshire west of London, the Cité Scientifique in southern Paris). In other cases, more peripheral areas have been able to attract large research centres and large production establishments. In the Nice region, there are IBM and Texas Instruments plants as well as many government research laboratories. Sometimes small firms settle or develop in regions without an industrial tradition, as in the neighbourhood of Aix-en-Provence or even more strikingly in Ticino. Such regions are singled out not because they offer a qualified labour market or an improved training system, but because they seem to be able to attract men, firms and capital.

Polarisation

During periods of the emergence of new technologies, spatially-polarised forms can appear whereby a great number of small firms develop as a result of a stimulus provided by a large plant or institution. The revival of SMEs since the early 1970s is in part associated with territorial concentration around units which have a strategic role, namely "knowledge centres". Silicon Valley, like the Cambridge region, is the result of a spin-off of firms, directly or indirectly, from a major local university. This happens when new technology breaks away completely from former trajectories so that the relationship with a research centre producing new knowledge becomes a determining factor. Industrial development of this kind does not really stem from existing or anticipated market demand, or from any existing firm; rather it results from scientific knowledge, and relies on new small firms which are uniquely capable of taking major risks in the form of radical innovation. Of course, as noted earlier, in reality this kind of polarization based on concentrations of knowledge probably also reflects other development processes. Thus in California, the presence of an enormous military demand has provided a solid basis upon which the development of the Silicon Valley has been based.

Many mechanisms unite in the formation of an industrial complex: the capacity to create clusters of activities round a large knowledge centre and some leading firms, the gradual specialisation of small firms which are encouraged to associate in order to launch new products, the formation of large firms which help to organize the industrial environment and spur on its development. Depending on the importance of these various mechanisms, the complex will remain centred on knowledge and research (Cambridge, Grenoble-Meylan) or will be able to develop into a more complete industrial centre (research, production, markets), as for example with Silicon Valley. Thus to some extent at least, large aviation plants have given rise to the beginning of a polarised concentration development in Toulouse (Gregoris, 1986) as well as in Bristol (Boddy and Lovering, 1986). However, Boddy and Lovering clearly show that the formation of a true industrial complex experiencing cumulative development implies a strong sectoral and technological unity which is not present in Bristol. In addition, generalized interaction, characteristic of Silicon Valley, is by no means compatible with the structure found in Bristol. In the latter case, large firms and large government research centres internalise the processes and do not encourage the spin-off of new firms in the area.

As a general conclusion, then, it is vital to distinguish between clusters of high-technology activity which have developed by the process of <u>attraction</u> (numerous firms attracted to an area for the same reasons), and those which are <u>polarised</u> (firms creating by their development and the links they have established between each other opportunities for further new firms to appear). From this perspective, "Silicon Glen" in Scotland is, on the whole, simply a concentration of American electronics firms which were looking for an English-speaking location which would allow them to supply the EEC market, and at the same time receive a good deal of public aid. Likewise in Bristol, there is a marked job-concentration related to defence, located in that area for historical reasons associated with the success of certain products developed locally in plants which now form part of British Aerospace. Bristol's aviation enterprises are responsible for the formation of a local market which also helps to explain the development of new firms in that area. In the south Paris suburbs, the local market for scientific instruments accounts for some of the high-technology development which has occurred in recent years. While inter-enterprise relations and particularly sub-contracting do seem to be flourishing, this is even more true of the Paris region as a whole than inside the southern suburbs alone. Thus the spatial extent of local synergy here does not always coincide with the limits of areas in which high-technology firms are concentrated. In each of these examples, a different form of attraction is found as well as a different combination of attraction and polarisation.

Table 2.2 attempts to provide a synthesis of the different mechanisms involved in the development of high-technology complexes.

Table 2.2 Mechanisms Underlying the Development of High-Technology Complexes

1. The initial establishment of a complex (the origin of firms)	
- location through government decisions	Sophia Antipolis
- attraction through government incentives	Scotland
- attraction through the image of the area	Sophia, Phoenix, Austin
- attraction through local demand for high-technology products	South Paris
- internal spin off	
- from knowledge centres	Cambridge
- from local firms	Silicon Valley
- high rate of internal firm creation	
- local socio-professional adapted structure	Aix-en-Provence
- local profit opportunities	Silicon Valley

Table 2.2. (cont.)

2. The operation and growth of a complex	
- local customers	Bristol, South Paris
- local suppliers	Milan North East
- sub-contracting on a local basis	Silicon Valley
- local inter-firm co-operation	Silicon Valley
- local scientific/technical links	Silicon Valley

A TYPOLOGY OF INNOVATION AREAS

On the basis of the categories and different mechanisms discussed above, table 2.3 attempts to classify technologically-innovative areas, on the assumption that most areas are probably combinations of the various ideal-types identified earlier.

Technological innovation is not a homogeneous category. Not only is it possible to distinguish, as is normal, product innovations from process innovations, not only is it necessary to treat separately innovations in traditional industries and those underlying the development of high-technology sectors, but it is also important to distinguish between the completely innovative or the partially composite nature of different technologies. Sometimes, products which are the result of new technologies are completely new; their conception, their manufacturing methods, the equipment employed, their use and their markets are all new. In other cases, however, the equipment required is already known, and manufacturing blends old processes (mechanical or micro-mechanical, for example) with new ones. Markets already exist, trademarks maintain their value. Following from this, it can be argued that the observed types of technologically-innovative areas can also be categorized by their capacity to adjust to one or other of these innovation types. When products and processes are almost all new, it is not surprising that the area which serves as a breeding ground for new activities has no industrial past, as this would not be of any use. However, if some part of former know-how can be re-utilized, or if local market demand remains, then areas which were once able to

Table 2.3 Towards a Typology of Technologically-Innovative Areas

		Regions of no industrial tradition	Regions with industrial tradition
Labour market		In formation	Changing
Local dynamism	Development THROUGH DEMAND	government demand defence spending research (Silicon Valley, Orange county)	BRANCHING concentration of innovation capacity (South Paris) sub-contracting, demand for technologically-advanced products
	Development from SUPPLY CAPACITY	SUPPLY OF KNOWLEDGE. POLARISATION spin-off from knowledge centres (Cambridge, Route 128)	PRESENCE OF INDUSTRIAL MILIEU AND KNOW-HOW. BREAKING-BRANCHING - adjustment of firms - spin-off from disappearing firms (Besançon) recycling of local assets, reconversion
External dynamism	LARGE FIRMS	ATTRACTION supply of location factors location of large firms technological and R and D centres (Sophia Antipolis)	ATTRACTION-JUXTAPOSITION supply of location factors location of production plants next to existing industrial system in decline duality (Charleroi, Newcastle, Valenciennes)
	SMALL FIRMS	Supply of location and "atmosphere" factors location of innovative SMEs (Ticino, Aix-en-Provence)	

create an efficient industrial milieu may, given time for adjustment, be able to develop some capacity for innovation.

This suggests a new approach to deciphering an area's capacity for innovation, based on a distinction between those areas which can start from scratch, those which are able to achieve a synthesis between an outdated activity and new techniques which are capable of revitalizing it, and, finally, those whose industries are not capable of developing or absorbing new technologies, and where such a synthesis is therefore not possible. The creation of Silicon Valley in a previously totally-unindustrialized area thus reflected the fact that semiconductor production did not involve recycling existing technology and know-how, whereas reconversion of the Swiss Jura was possible because the key innovation here, the electronic watch, required familiarity with the market, the commercial milieu and with some of the labour skills of the region. Conversely, when the know-how associated with old activities is of no relevance to new technologies, the regions in question have no alternative open to them other than to wait for the advent of external firms (Charleroi, Newcastle, Valenciennois).

CONCLUSIONS: TOWARDS A DYNAMIC APPROACH TO UNDERSTANDING INNOVATIVE AREAS

A typology such as the one above compares different kinds of milieus without regard for their historical development, as if they were all at the same stage of their evolution. To a certain extent, it would seem that the capacity of local environments to absorb technical progress depends less on their particular characteristics than on the point of development they have reached. A region which is not yet industrial may become innovative if a random event initiates innovation, and a process of polarization is set in motion. Its characteristics as well as its forms of technological development have nothing in common with what can be seen in former industrial milieus dependent on a given technical sectoral basis, and which are suddenly subject to the impact of a new technology. In other words, situations displayed in the cells of table 2.3 above characterize not alternative processes, but often successive ones.

Entirely new technologies are inevitably initially localized in their occurrence, as with the establishment of new high-technology industries dependent on government expenditure in previously unindustrialized areas. Alternatively, some large research or production establishments may be set up in non-industrial areas as a result of government decisions (aviation decentralised to Toulouse before World War II, large public research centres established at Sophia Antipolis in the 1970s, aviation plants in Bristol). Once this initial stage is over, technological development will depend on the area's ability to create a process of polarisation around some central focus. This may be either the large firm which was originally established there, and whose influence spreads throughout the neighbouring area thus inducing industrial development (SNIAS in Toulouse, IBM in Montpellier...), or a large knowledge centre which will provide scientific and technical innovations as well as potential entrepreneurs, engineers, and a market for scientific instruments (Paris). As it reaches maturity, the complex will, sooner or later, have to face a totally different problem: the arrival of a new technology, different from that which had been the basis of initial development of the area. It would be interesting to see what is the situation in, for instance, the Paris area which remains the leading industrial region in France. The many assets that the region possesses (university research, proximity to decision-making centres, a powerful and diversified industrial milieu) bring about a capacity for innovation which manifests itself in parts of the region where sites are available close to existing industrial complexes, yet which are sufficiently far away to facilitate the formation of a new labour market.

An industrial milieu hit later on by the arrival of a new technology faces an alternative; if it is able to re-employ its assets and effect a synthesis between the new techniques and its own know-how, it will initiate the breaking-branching process described above. Otherwise, it will have to attract firms from outside by publicizing the locational advantages it possesses (attraction/juxtaposition). If it does not manage to do either, the area will be left to suffer economic decline in the absence of technological innovation.

The mechanisms underlying technological development of areas are historically specific. Let us look for example at Silicon Valley. Its preliminary development until the beginning of the 1950s relied, to a large extent, on individual initiative, on the leading role of certain individuals. Key events here included the founding of the university by Leland Stanford, the initiatives of Terman and the creation of Stanford's Science Park, the attraction of capital destined for military purposes due to the presence of such men as Teller, the arrival of Shockley, and the creation of a small firm in a garage just before the World War II by Hewlett and Packard. From the 1950s onwards, another process took over: the development of a strong defence demand adapted to the first stage of the development of a new sector. In the 1960s, the polarisation process began, based on the spread of semiconductor development which was this time oriented to use in micro-electronics for a private market. This process broke through the limits imposed by the previous growing domination of large firms and the spatial dispersion which had begun in the 1960s. Today, there is a new marked specialisation in terms of function at the stages of research, development and overall management. But the reputation of Silicon Valley has continued to attract firms and firm founders from outside which have reinforced the purely local dynamism. The success of the region over the last 50 years has thus been based on a chain of events. The achievements of the area today depend on a succession of development processes rather than on the right functioning of one or the other.

The shift from one process of territorial technological innovation to another of course also reflects changes in the nature of the production system, and sectoral restructuring with changing market demand. In the Silicon Valley case, for example, the shift from polarization to a global pattern of production based on the international division of labour reflected increasing large firm concentration and a shift to mass-production of electronic components. Equally, the continuing development of electronics firms in Silicon Valley and nearby newer complexes such as Santa Cruz county reflects the vital contemporary influence of linkages between, and spin-off of, SMEs in the context of growing demand for "customized" semiconductors, a development which is linked to increased competition in standardized components.

All these factors, of changing market demand, the role of local or external capital and enterprise, the adaptability of existing know-how and labourforce skills, and the nature of current technological opportunities, thus help to shape the evolution of industrial milieus confronted by new technologies, and the degree of success or failure which they achieve in this respect.

REFERENCES

Aydalot, P. (ed.) (1986) Milieux innovateurs en Europe, GREMI, Paris

Aydalot, P. (1987) The role of small and medium size enterprises in regional development: conclusions drawn from recent surveys. In Giaoutzis, M. and Nijkamp, P. (eds) Small and medium size firms and regional development, Croom Helm, London

Boddy, M. and Lovering, J. (1986) High technology industry in the Bristol sub-region: the aerospace/defence nexus. Regional Studies, 20, 3, pp. 217-31

Castells, M. (1984) Towards the informational city? High technology, economic change and spatial structure. Institute of Urban and Regional Development, University of California, Berkeley, Working Paper 430

Decoster, E. and Tabariés, M. (1986) L'innovation dans un pôle scientifique et technologique; le cas de la cité scientifique Ile de France sud. In P. Aydalot (ed.), Milieux innovateurs en Europe, GREMI, Paris, pp. 79-100

Gordon, R. and Kimball, L.M. (1986) Industrial structure and the changing global dynamics of location in high technology industry. Silicon Valley Research Group, University of California, Santa Cruz, Working Paper 3

Grégoris, M-Th. (1986) Le technopôle Toulousain: un terme moderniste pour qualifier un processus ancien. In Association de Science Régionale de Langue Française, Technologies nouvelles et développement régionale, Centre Economie-Espace-Environnement, University of Paris 1, pp. 180-7

Hoogteijling, E., Gunning, J.W. and Nijkamp, P. (1986) Spatial dimensions of innovation and employment: some Dutch results. In P. Nijkamp (ed.), Technological change, employment and spatial dynamics, Springer-Verlag, Berlin, pp. 221-43

Planque, B. (1985) Le développement par les activités à haute technologie et ses repercutions spatiales: l'exemple de la Silicon Valley. Revue d'Economie Régionale et Urbaine, 3

Pottier, C. and Touati, P-Y. (1986) Les conditions de l'innovation dans les régions d'industrialisation ancienne; le case de Besançon. In P. Aydalot (ed.), Milieux innovateurs en Europe, GREMI, Paris, pp. 247-66

Rothwell, R. and Zegveld, W. (1982) Innovation and the small and medium sized firm, Frances Pinter, London

Scott, A.J. and Angel, D.P. (1987) The US semiconductor industry: a locational analysis. Environment and Planning A, 19, 7, pp. 875-912

Viola, M-C. (1986) Création d'entreprises et innovation dans une région de vieille tradition industrielle: la capacité du Valenciennois à se reconvertir, Memoire de D.E.A., Novembre 1986, Université de Paris 1

Chapter 3

Functional Integration and Locational Shifts in New Technology Industry

Roberto Camagni

INTRODUCTION

During the sixties and seventies a clearcut reversal of previous spatial trends took place in all advanced economies, including those of western Europe, with the emergence of what was called "non-metropolitan industrialisation". Intermediate and peripheral regions and rural areas within advanced regions experienced sustained growth, in sharp contrast to the slowdown and even crisis of formerly "strong" metropolitan areas.

Sometimes this growth was the direct result of a decentralisation strategy away from central areas, but in most cases it was the outcome of indigenous potential and "development from below" (Camagni and Cappellin, 1981; Stohr, 1986). The basic elements of this "diffused" growth pattern were: high flexibility of the production mix attained through a "system" of small specialised enterprises; relaxed industrial relations and a new social consensus; a high rate of innovation diffusion throughout the "system" areas; and strong spatial synergies in production organisation, finance and sales (Brusco, 1982; Piore and Sabel, 1984; Fua' and Zacchia, 1984).

In contrast to a host of local success stories, not just in the "Third Italy" but in almost all advanced countries, stood the crisis both of large firms and the most urbanised and industrialised areas (Aydalot, 1984), which were not able to develop a successful response to a changing and turbulent environment.

This scenario, about which there exists widespread agreement among contemporary scholars, may well be destined to change somewhat in the next decade. This is because powerful new elements have recently come into play, which are likely to help the economic redeployment of those areas which boast long-standing industrial cultures and in particular large, service-endowed, metropolitan areas. These elements concern the strategic, organisational, and therefore spatial effects of new information technologies on industries, both those producing new high-technology devices and those applying them to the production process. Together, these two categories may be termed "new technology industry".

Some of these elements are already apparent and have determined a new shift in the locational pattern of such industry. Others on the other hand may only be forecast on the basis of "weak signals" and through the more influential prescriptions of business management literature.

The aim of this chapter is to highlight the nature of these new elements, which mainly concern the behaviour of large and medium-sized firms, and to discuss their likely impact on (micro-economic) location strategy and the (macro-economic) development of spatial disparities, with particular reference to western Europe.

Apart from the process of internationalisation (Amin and Goddard, 1986), which will not be considered here, there are three key words in this new order: innovation (and in particular, "continuing innovation"), information (and "informatisation" of all aspects of the firm's structure), and functional integration. By means of these elements, old industrial structures are challenging the innovativeness, flexibility and spatial synergies of small enterprises and their spatial systems.

THE OBJECTIVE: CONTINUING INNOVATION

An important element has become clearly visible within the new technology industries. The traditional product life-cycle has lost its well known shape and pace, becoming shorter (diminishing from about 20 years to five years) and truncated after the development phase ("losing its tail") (Oakey, 1985: Camagni and Rabellotti, 1986). The effects of this are twofold. First of all, the firm is condemned to constant innovation as such things as "cash cow" products are becoming more and more rare. Secondly, the scope for a strategy of selective decentralisation of mature production towards less developed areas is diminishing.

A good example of this trend is provided by the share of total products sold by Siemens that are accounted for by products belonging to different periods. Products developed more than ten years before accounted for 32% of total sales in 1969, 25% in 1977 and 19% in 1983; on the other hand, products developed in only the last five years accounted for 38% in 1969, 45% in 1977 and 53% in 1983 (Bullinger et al., 1986).

New technologies may substantially help this process of the speeding up of product innovation, partly by increasing the productivity of the creative process, as through computer-aided design, and partly by supplying the opportunity of product "rejuvenation" through more advanced and reliable production processes (Camagni, 1986a).

In general, from an aggregate and macro-economic point of view, after the productivity slowdown of the last decade, an upswing of innovative processes and productivity increases has undoubtedly occurred during the 1980s, a fact that long-wave theoreticians may well link to the appearance of a new cycle, based on the "information technology paradigm" (Freeman and Soete, 1986).

THE INSTRUMENT: NEW INFORMATION TECHNOLOGIES

By information technologies we mean all hard and soft devices which allow the storage, transmission and manipulation of information. The sectors involved are mainly five: semiconductor production, computers, telecommunication equipment, production of flexible automation and process control devices, and software for office and factory automation.

The automation technology paradigm affords wide-ranging opportunities for supplying a new range of products and product innovations both to society and to the economic system, once an efficient link and a good "matching" is achieved between new technological opportunities and the socio-institutional sphere. What is more interesting for our purpose, it presents a new range of advanced production devices and process innovations that pervasively enhance the productivity, competitiveness, and profitability of industry. In fact, it is increasingly <u>integrating</u> different activities within the manufacturing system from design to the market, through a sophisticated information network.

New technologies of adaptive controls and feedbacks, electromechanical sensors and pattern recognition, for example, require a substantial amount of "tool point" information. But at the same time they help to create it. Once the process is fully integrated, "information usage accelerates, as does potential for improved control, and it becomes more feasible to collect data, forecast and control, schedule and optimize" (Jelinek and Golhar, 1983).

Robotised and flexible manufacturing systems in particular permit:

- a considerable reduction in direct labour,
- an appreciable reduction in unit capital costs,
- a breakdown of the traditional trade-off between efficiency and variety of products manufactured, allowing product differentiation at zero marginal costs within the same plants,
- the optimisation of internal flow of components and parts,
- a substantial reduction of idle times for both machines and components,
- an appreciable reduction of floorspace needed in production plants (30% on average)
- the possibility of unmanned production,
- higher quality of the product and time reliability of the process,
- enhanced managerial control over the production cycle (Camagni, 1986b).

Flexible automation systems share a common feature with computers and office automation processes generally: their efficiency is closely dependent upon the general adoption philosophy. In particular, during a common early stage in which the new technologies are mainly "substituting" for the old ones in order to perform the same tasks, the productivity of the process may rise but it soon reaches an upper threshold. A second stage is then required, which may be called the integration stage, in which the new technology is used to perform new tasks, through the integration of different machines, different operations, and different functions.

In the case of computers, we pass from operational tasks (like pay computing) to managerial ones like control and forecasting. In the case of flexible automation systems, we pass from the adoption of "stand-alone" robots and numerical control machines to systems of machines served by robots ("manufacturing cells" and "flexible manufacturing systems"). The new philosophy needs a general reconsideration of the production process, with the redesign of the layout and of the form of the components, in order to permit blind mechanical handling.

In this later case, a third stage may be defined, in which the process of integration reaches a higher level, linking the factory system with the managerial one, and creating synergies between production, engineering, marketing and strategic planning (Fig. 3.1.)

Figure 3.1: Stages of Flexible Automation Adoption

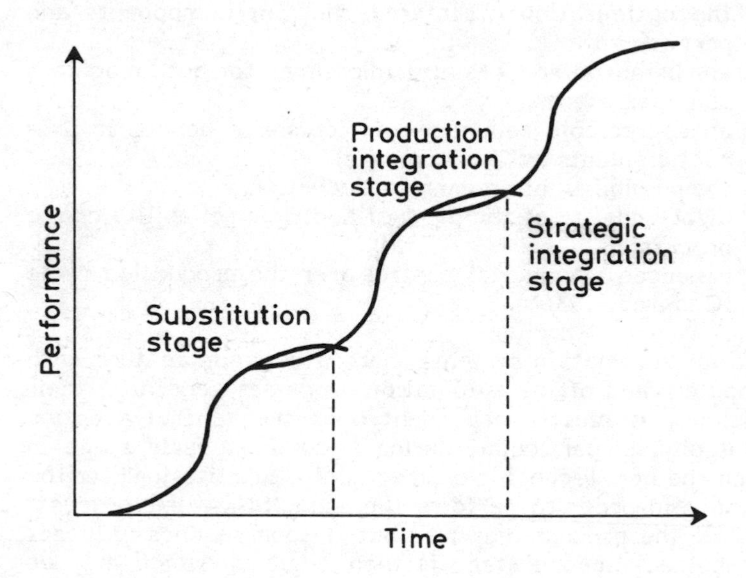

Source: Camagni (1986)

THE METHOD: FUNCTIONAL INTEGRATION

It has long been recognized by organisation scholars that the profound functional specialisation of the big enterprise, designed to achieve economies of scale and higher professional know-how, presents the risks of internal segmentation and bureaucratisation, and in particular a loss in terms of efficient exploitation of information arising from every day operations in each department.

This risk is particularly evident with respect to innovation activity, as a new product has to be at the same time marketable, easy to manufacture at low costs, and coherent with the firm's general strategy. What is needed is therefore "an organisation which provides collaboration between scientific innovators and sales and production specialists, so that: a) the skills of the innovators can be directed at market needs and technological problems; b) sales and production specialists can be actively involved in the commercialisation of ideas developed in the laboratory; and c) as a result, ideas can be transferred smoothly from laboratory prototype to commercial reality" (Lorsch and Lawrence, 1968).

It is not sufficient to build up a huge R&D department to guarantee a fast flow of innovations, as industrial economists have recognised; this department has also to work in direct contact with other strategic departments such as marketing and production. The integration between marketing, production and R&D is the first type of functional integration suggested by the literature, through permanent crossfunctional co-ordinating committees (Lawrence and Lorsch, 1969) or through "ad hoc structures" with ever shifting internal relationships, built up on ad hoc projects (Mintzberg, 1983).

Along the same lines, J. R. Galbraith, the early proposer of complex "matrix" organisations, has recently argued that organisations designed for normal, routine, tasks must be clearly distinguished and differentiated from organisations addressed to innovative objectives. In other words, an organisation designed to produce the millionth piece or to perform the millionth operation is not suitable for executing something for the first time! He then suggests that integrated structures addressed to the development of new ideas should be financially, organisationally and physically separated from day to day structures, and organised in what he calls "reservations" (Galbraith, 1983).

A second kind of functional integration, at the strategic level, suggests the involvement of production managers into the strategic planning process, in contrast with the traditional view that considers them as stepchildren to financial or marketing people. This approach is not in itself novel (Skinner, 1969), but has gained new significance in recent years with the emergence of flexible automation systems. These devices in fact refute the common theory of a unique path of process development, from fluid processes in the early stages of the product life-cycle to systematic, extremely rigid and capital intensive processes in the mature phase. Previous manufacturing experience saw progress in the building of special purpose, "dedicated" mass production machines, which involved a clearcut trade-off between efficiency and flexibility. In contrast, the new flexible systems permit the production of a range of products with the same facility, and call for a strategic exploitation of this opportunity.

Of course, this exploitation "entails new customers, new channels of distribution, new methods of selling, ..., and a new vision of the firm and its strategic mission" (Jelinek and Golhar, 1983). It also requires a new strategic integration of engineering know-how with marketing and long-term planning.

A third form of functional integration, which follows from the preceding ones, concerns the integration between product and process innovation functions. This is an increasing phenomenon in all industrial countries. In Italy, for example, a large inquiry carried out by the national statistical office on innovation diffusion in manufacturing industry found that this integration was the most frequent behaviour among innovative firms (53% of cases, over more than 16,000 innovators). Very similar results were also attained in the GREMI inquiry in the Swiss Jura (42%) (Istat, 1986: Maillat and Vasserot, 1986).

In fact, the objective of a speedy introduction of a competitive product on the market is increasingly reached by forcing engineers, applied researchers and marketing personnel to work together in "mission" units from the conception phase to industrialisation and production, on an efficient (and not just a prototype) scale. This is particularly valid in the new technology industries, where the product life-cycle is shorter and the need for continuous innovation stringent (Camagni and Rabellotti, 1986). In these cases, it may be argued that the traditional Abernathy and Utterback curve of process innovation is forced into following strictly the product innovation curve, as is shown in Fig. 3.2.

Figure 3.2: Innovation Rates along the Product Life-Cycle

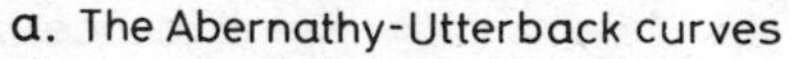

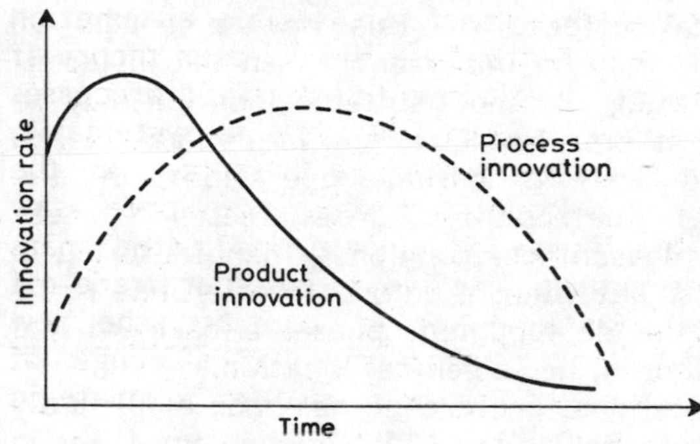

b. The high-technology industry curves

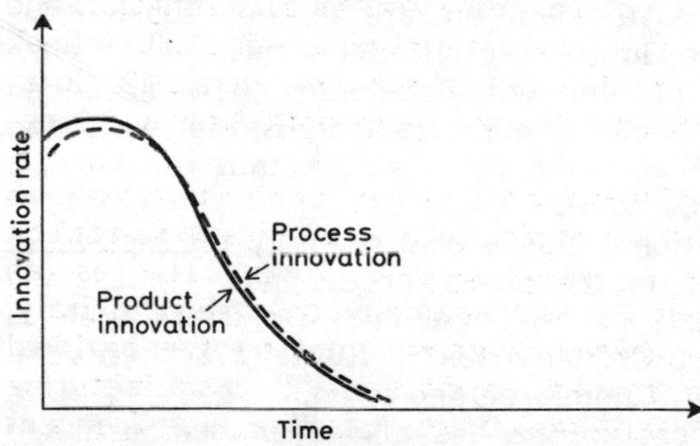

Source: Camagni and Rabellotti (1986)

NEW FIRM STRATEGIES AND NEW LOCATIONAL PREFERENCES

A new philosophy or rather a new firm strategy follows from the preceding elements in the case of new technology industries. By new technology industries we mean not just high-technology sectors, but all production processes that make wide use of new information technologies. In this section of the chapter the different elements of this new strategy are reviewed (x), together with their possible spatial consequences (x').

a) **Functional integration versus functional specialisation.** As we have seen, operational efficiency requires specialisation and broad operational volumes, while innovative efficiency requires synergism between functions and the physical integration of product and process innovation specialists.

a') The spatial effects of this new emphasis on functional synergism may be important. When the different functions are separated, the most suitable location may be found for each of them: headquarters in central cities, administration and routine bureau operations in the hinterland of large metropolitan areas, research near universities and in pleasant environments, manufacturing in the peripheral areas. But when a unique location has to be found at least in the development phase, a baricentrical location has to be chosen, i.e. a central location.

This kind of locational preference has been empirically demonstrated for "mission" units of high-technology firms in the Paris and Milan areas (Decoster and Tabariés, 1986: Camagni and Rabellotti, 1986) and for the new subsidiaries or firms created by large diversified concerns when confronted with the problem of developing new advanced products in biotechnology and electronics, as with the Belgian experience analysed by Alaluf within the GREMI inquiry (Alaluf et al., 1986).

The same locational bias is confirmed by the high level of accessibility to metropolitan areas, universities and international airports of high-technology clusters such as Santa Clara county, Orange county, Route 128, Berkshire and the M4 Corridor, Cambridge and so on.

But the same preference has also been demonstrated for large firms with an extensive and innovative adoption of information technologies. For example, the production of Fiat's FIRE 1000 engine, deliberately designed in order to allow completely robotised assembly and therefore carefully studied for four years in co-operation with flexible automation specialists, engineering consultants and the Turin Polytechnic, was not just designed but first put into operation in a small automated plant near Turin. Only after this was production rapidly serialised in a much larger plant in Southern Italy, in the Abruzzi region.

b) **Flexibility versus rigidity**, both in capital equipment and in the use of the labourforce. The new flexibility property of robotised systems has already been illustrated, but the increased dynamism and the informal organisational structure of new technology firms require new management attitudes with respect to labour organisation and industrial relations.

In the case of this last aspect, very instructive practical experience comes from U.S. high-technology firms and particularly from Japanese firms. Through their European subsidiaries they have adopted new labour organisation practices, more and more accepted by other indigenous firms, like: "active consent" strategies with the institutionalisation of labour flexibility in written clauses and "no strike" agreements on the one side, relative job security, single worker status and the creation of quality circles on the other; partial rejection of the taylorist wisdom with loss of precise job description and simplified pay structures, based on behavioural skill rather than on piece-work bonus systems; and direct contact between managers and the shopfloor and workers' identification with the company (Morgan, 1984).

b') The spatial consequences of these elements are mainly indirect, in that they bridge the gap between small and large firms from the point of view of flexibility and labour consent. Large firms - which, contrary to what is commonly believed, are also dominant in high-technology sectors (in the U.S.A. 89% of employment in high-technology sectors is in firms with more than 100 employees: Armington, 1984) - may reduce the comparative disadvantage that has been often cited as one of the major causes of the relative crisis of "central" areas through these practices.

c) **Customisation versus product standardisation**. One of the main new elements which has developed since the 1960s on the demand side has been the increasing shift in consumer demand away from mass products in favour of differentiated, personalised articles. Demand for variety has increasingly replaced demand for cheap standardised products, and this has created wider scope for the development of small, specialised firms (Boeri, 1986). Large firms were traditionally not in a position to capture all segments of a differentiated market, even by sub-contracting special production to small firms. Now, new technologies supply them with a powerful tool for product differentiation at low marginal costs.

c') The spatial consequences of this last element are once again indirect, and reflect the increasing market share of large firms, producing "customized" products through flexible automation systems and the parallel reduction of market "niches" for (dispersed) small firms.

d) Economic literature is increasingly defining this new comparative advantage made possible by new technologies as **exploitation of economies of scope versus traditional economies of scale**. Economies of "scope" are those economies of joint production resulting from the use of a single set of facilities to produce, or process, more than one product (Teece, 1980: Chandler, 1986).

d') In order to ascertain the possibilities for small enterprises also to adopt new technology, and therefore the possibility of them catching up large automated firms in this respect, we must recall that the exploitation of economies of scope requires a high volume of total, even if differentiated, production. This is because flexible automation systems imply an investment in hard and soft devices which is comparable with that of previous dedicated transfer lines. This important element conflicts with the common belief that new technologies are also easily adopted by small enterprises.

e) **"Just-in-time" input organisation versus investment in inventory**. The extended planning capacity (and necessity) of the production flow made possible by automation systems renders the Japanese "kanban" or "just-in-time" principle possible, thus permitting a substantial reduction of working capital.

e') This practice requires a new, tighter organisation of the location of suppliers, who, in the case of Japanese experience in the car industry (Toyota), are explicitly required to locate at less than one hundred kilometres from the assembly point (Lambooy, 1986). In contrast to previous theories about the necessity for a "world car", developed at MIT, according to which only a few assemblers and a few specialised parts producers were likely to survive in world competition, many smaller producers have been able to remain profitable, buying parts from a system of closely located and directly controlled suppliers and subcontractors. This pattern is perhaps becoming more widespread than the former at an international scale, but probably more concentrated at the interregional scale.

A good example of the same process is provided by the experience of the European subsidiary of Caterpillar, located in the Charleroi region, which is increasingly substituting imports of parts from the U.S. by purchases from a host of local suppliers, often stimulated and technologically sustained by the firm (Alaluf, 1986).

f) **Concentration of strategic and planning power and transfer of intermediate responsibility at the plant level.** New technologies permit and require a concentration of strategic planning and sometimes of production planning, but they also allow the possibility of delegating the development of specific products to peripheral areas.

f') Once again, the possibilities of transferring responsibility and decision-making apply at the international level, new research and planning responsibilities being capable of decentralisation to overseas subsidiaries, located in service-endowed metropolitan areas. At the interregional level, however, the same mechanism does not seem likely to operate, due to lack of sufficient know-how in the peripheral areas. In addition, because of the diminishing need for an unskilled labourforce, branch plants are increasingly located close to markets rather than close to cheap labour areas.

g) **External collaboration for innovation versus internal development.** Increasing difficulty in basic research work and innovative processes force firms to search for international co-operation and joint venture possibilities, together with co-operation with local public and private research and consulting institutions. As Perrin has cleverly put it, "on observe une transition des pratiques passées de generation interne des strategies d'innovation a des strategies selectives de controle des marches innovateurs" (Perrin, 1986).

g') It is easy to grasp the revitalisation possibilities for highly urbanized areas and for highly cultured and "creative" environments which are implicit in the previous statements.

h) **Continuing innovation and declining product maturity.** As already stated, mature products are now readily substituted by new ones. Because of this, traditional expectations of production decentralisation to less-developed areas during later stages of the product life-cycle may be profoundly altered, or even reversed. Less scope remains in fact for international decentralisation of mature production.

For example, in the mechanical engineering and machine tool industry, a sector where some new industrialising countries were in the 1970s developing some initial degree of expertise and know-how, new technologies and product innovations like numerical control and computer-aided design are re-establishing the leadership of advanced countries (Erber, 1986), thus halting the previous trend.

i) **Decreasing scope for international selective decentralisation of production phases.** The pervasive character of technological advances is going powerfully to affect all sectors and also all phases of the production process. Other industries, in addition to traditional labour-intensive activities such as textiles and clothing, may also expand successfully in new locations in advanced countries, if "rejuvenated" by advanced process technologies and supplemented by advanced marketing and organisation (the Italian case is typical in this respect). Moreover, single production phases like assembly, previously untouched by automation and therefore decentralised at an international scale, may be recalled to "central" areas due to:

- new advanced automation possibilities,
- lower incidence of direct labour costs,
- decreasing need for floorspace in manufacturing production
- greater need for skilled labour and external services.

In fact, these considerations, taken together, reduce or even annihilate the comparative advantage of less developed areas with respect to advanced ones. The use of advanced process technologies is by no means simple, and empirical studies have demonstrated the "conservative" nature of their spatial diffusion process (Camagni, 1986a; 1986b).

An interesting example of this trend is given by the reversal that is taking place in the international location strategy of assembly plants in the semi-conductor industry, due to new automation possibilities. In recent years several major U.S. manufacturers have constructed or are currently establishing highly automated VLSI assembly lines onshore in the United States. Examples here include Motorola (in Arizona), Intel (Arizona), Fairchild (Maine), and Applied Microcircuits (California) (Ernst, 1986).

CONCLUSIONS

In the past, regional disparities have mainly evolved under the influence of two main factors: a long-term, "entropic" trend towards increasing spatial homogeneity of know-how, infrastructure endowment and quality of production factors, and a more complicated medium-term cyclical factor, generally involving increasing concentration during the upswings, but increasing diffusion during the later stages of the cycle. During the 1960s and 1970s, these two elements worked together in favour of the periphery, which was able to "catch up" with respect to the "centre" through imitative innovation processes and original organizational innovations. This view reflects in particular the success of "systems areas" of small specialized firms, based on the creation of a new social consent and spatial synergies.

The emergence of what is already termed the "fifth kondratief" cycle, based around the new information technology paradigm, is destined in our view to reverse some of the preceding trends, at least as far as the initial stage of the cycle is concerned (about 15 years). This is because, as we have argued in this chapter, adoption of these technologies is by no means simple, and requires high levels of skill and well-developed urban environments endowed with advanced services. New technology industries are thus always geographically more concentrated than other industrial sectors, as was also demonstrated by the GREMI inquiry on the Netherlands (Lambooy, 1986b). Moreover, aided by these technologies, large firms are increasingly finding the right answer to cope with the turbulent economic environment, overcoming their previous widespread managerial and social crisis.

New strategies for increasing flexibility of the production process and of labourforce utilisation, new management/labour relations and social consent, and the search for new synergies, both internal (through functional integration) and external (through expanded control over what may be called the "socio-economic network") are likely greatly to reduce the previous gap in relative locational advantage between central and peripheral areas, in favour of the former.

These elements may not perhaps change the overall trend in advanced European countries towards greater spatial homogenisation, but will nevertheless supply old industrial and central areas with a new powerful potential for revitalisation.

REFERENCES

Alaluf, M., Martinez, E. and Vanheerswinghels (1986) Situation économique, facteurs de redéploiement et innovation technologique: le cas de Charleroi. In P. Aydalot (ed.), Milieux innovateurs en Europe, GREMI, Paris, pp. 163-94

Amin, A. and Goddard, J. (eds) (1986) Technological change, industrial restructuring and regional development. Allen & Unwin, London

Armington, C. (1984) The changing geography of high technology business, Applied Systems Institute Inc., Washington D.C. (mimeo)

Aydalot, P. (ed.) (1984) Crise et Espace, Economica, Paris

Aydalot, P. (ed.) (1986) Milieux innovateurs en Europe, GREMI, Paris

Boeri, T. (1986) Modelling regional growth and spatial diffusion of innovation. Paper presented at the European Regional Science Congress, Krakow

Brusco, S. (1982) The Emilian model: productive decentralisation and social integration, Cambridge Journal of Economics, 6, 167-84

Bullinger, J., Warnecke, J. and Lentes, P. (1986) Towards the factory of the future. Paper presented at the EEC and Fast Seminar on New Production Systems, Turin, 2-4 July

Camagni, R. (1986a) Robotique industrielle et revitalisation du Nord-Ouest Italien. In J. Federwisch and H. Zoller (eds), Technologie nouvelle et ruptures régionales, Economica, Paris, pp. 59-80

Camagni, R. (1986b) The flexible automation trajectory: the Italian case. Paper presented at the International Conference on Innovation Diffusion, Venice, March

Camagni, R. and Cappellin, R. (1981) Policies for full employment and efficient utilisation of resources and new trends in European regional development, Lo spettatore internazionale, 2, pp. 99-135

Camagni, R. and Rabellotti, R. (1986) Innovation and territory: the Milan high tech and innovation field. In P. Aydalot (ed.), Milieux innovateurs en Europe, GREMI, Paris, pp. 101-28

Chandler, A. (1986) Scale and scope: the dynamics of industrial enterprise, (mimeo)

Decoster, E. and Tabariés, M. (1986) L'innovation dans un pôle scientifique et technologique: le cas de la cité scientifique Ile de France Sud. In P. Aydalot (ed.), Milieux innovateurs en Europe, GREMI, Paris, pp. 79-100

Erber, F.S. (1986) Patterns of development and the diffusion of technology. Paper presented at the International Conference on Innovation Diffusion, Venice, March

Ernst, D. (1986) Automation, employment and the third world - the case of the electronics industry. Paper presented at the International Conference on Innovation Diffusion, Venice, March

Freeman, C. and Soete, L. (1986) L'onda informatica, Edizioni del Sole 24 Ore, Milano

Fua', G. and Zacchia, G. (eds) (1984) Industrializzazione senza fratture, Il Mulino, Bologna

Galbraith, J.R. (1983) Organizzare per l'innovazione, Sviluppo e Organizzazione, 79, pp. 153-70

ISTAT (1986) Indagine sulla diffusione dell'innovazione tecnologica nellin'dustria manifatturiera italiana, Notiziario Istat, 41, June

Jelinek, M. and Golhar, J. D. (1983) The interface between strategy and manufacturing technology, Columbia Journal of World Business, Spring, pp. 26-36

Lambooy, J.G. (1986a) Information and Internationalisation: dynamics of the relations of small and medium sized enterprises in a network environment. Paper presented at the Round Table on Les PME innovatrices et leur environnement local et economique, Aix-en Provence, July

Lambooy, J.G. (1986b) Regional development trajectories and small enterprises: the case study of the Amsterdam region. In P. Aydalot (ed.), Milieux innovateurs en Europe, GREMI, Paris, pp. 57-78

Lawrence, P. and Lorsch, J. (1969) Organisation and environment: managing differentiation and integration, R. Irwin Inc., Homewood

Lorsch, J. and Lawrence, P. (1968) Organising for product innovation, Harvard Business Review, pp. 109-22

Maillat, D. and Vasserot, J.-Y. (1986) Les milieux innovateurs: le cas de l'Arc Jurassien suisse. In P. Aydalot (ed.), Milieux innovateurs en Europe, GREMI, Paris, pp. 217-46

Mintzberg, H. (1983) Designing effective organisations, Prentice-Hall, Englewood Cliffs

Morgan, K. (1984) Social innovation and the electronic industry: examples from South Wales and the English Sunbelt. Paper presented at the Urban Change and Conflict Conference, Sussex University, April

Oakey, R.P. (1985) High technology industries and agglomeration economies. In P. Hall and A. Markusen (eds), Silicon landscapes, Allen & Unwin, Boston

Perrin, J.C. (1986) Le Phénomène Sophia-Antipolis dans son environnement régional. In P. Aydalot (ed.), Milieux innovateurs en Europe, GREMI, Paris, pp. 283-302

Piore, M. and Sabel, C. (1984) The second industrial divide, Basic Books, New York

Skinner, W. (1969) Manufacturing: missing link in corporate strategy, Harvard Business Review, May-June, pp. 136-45

Stohr, W. (1986) Territorial innovation complexes. In P. Aydalot (ed.), Milieux innovateurs en Europe, GREMI, Paris, pp. 29-56

Teece, D. (1980) Economies of scope and scope of the enterprise, Journal of Economic Behaviour and Organisation, 1, pp. 223-47

Chapter 4

High-Technology Industry and Local Environments in the United Kingdom

David Keeble

TECHNOLOGICAL CHANGE AND INDUSTRIAL RESTRUCTURING

Since the mid-1970s, the United Kingdom in common with other European countries has been undergoing major changes in the nature, volume and location of manufacturing industry (Keeble, 1987). The most striking of these in aggregate terms was the acute decline of manufacturing output and employment with recession between 1979 and 1981 (see Figure 4.1), after earlier substantial growth of production in the 1960s and early 1970s. Since 1981, manufacturing recovery has been characterised by high rates of growth of labour productivity and output per head, though manufacturing employment has continued to fall.

The dramatic changes since 1979 portrayed in Figure 4.1 have been accompanied not only by massive redundancies and very high unemployment, but also by major shifts in industrial organisation. The latter include the rationalisation and modernisation of existing firms and industries, the widespread adoption of new work practices and production technologies, a relative and absolute growth in numbers of small and new firms, considerable expansion in high-technology production, and substantial foreign inward investment. This 'industrial restructuring' is viewed by Martin (1987) as reflecting the development of a new regime of **flexible** capitalist accumulation, different in kind from the previous **monopolistic** regime which operated from 1940 to 1970, and necessitated by the deepening crisis of profitability and uncompetitiveness which characterised British - and European - industry over the 1960-80 period. This "flexibilisation of accumulation and production" (Martin, 1987, 11) is associated with major social and political changes, and is impacting and interacting spatially so as to

Figure 4.1 United Kingdom Manufacturing Output, Employment and Output per Head, 1963–86

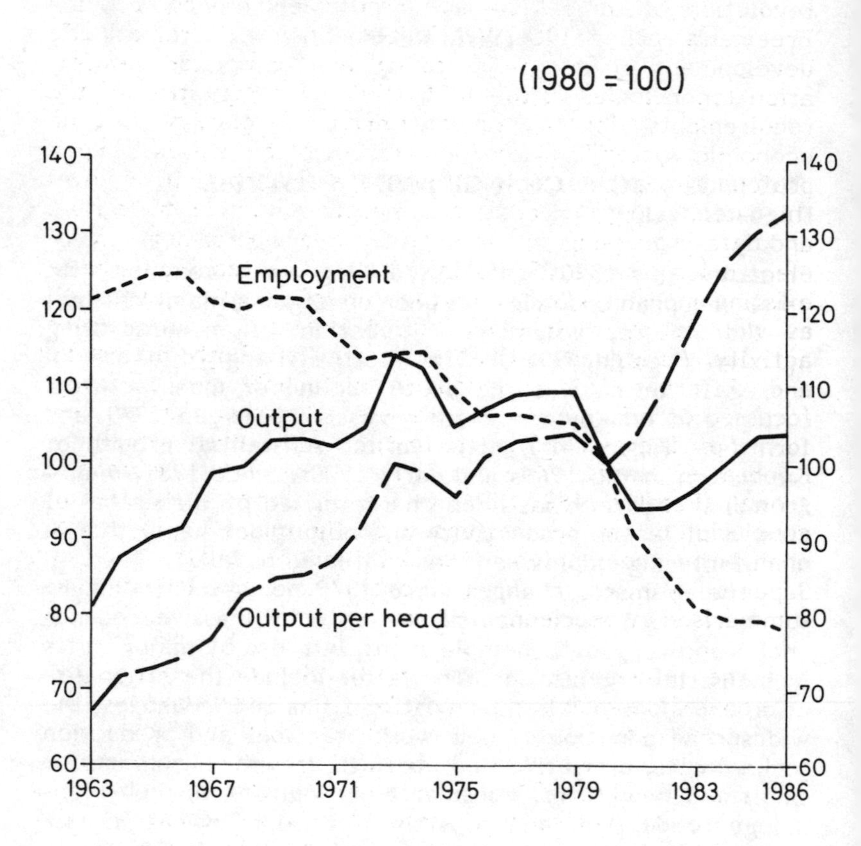

differentiate localities and regions in new ways.

The role of technological change and innovation in this process of contemporary industrial restructuring is clearly considerable, as Martin fully acknowledges. Indeed, commentators such as De Jonquieres (1987) argue that, "that Britain is today in the grip of a technological revolution of some kind is indisputable", a view echoing Freeman's earlier (1986, 109) judgement that the remarkable development of micro-electronics and associated information technologies in the 1970s and 1980s "satisfies all the requirements for a Schumpeterian revolution in the economic system". These new technologies are exerting a profound impact not only through the development of novel "high-technology" products and industries such as computers and data-processing, but through the adoption of new micro-electronics **process** technology by an ever-widening range of existing industries (cars, textiles, mechanical engineering), as well as by their effect on domestic life and leisure activity. As argued in Chapter 1, they are both distinctive and different from previous new technologies in being focussed on information and knowledge, rather than on new forms of energy, and seem to herald the onset of a new Kondratief long wave of innovation-related economic growth (Preston, 1987). The latter is indeed a fundamental conclusion of two recent reports on contemporary industrial change by the Confederation of British Industry and the Japanese Ministry of International Trade and Industry, as summarised by Searjeant (1986):

> The information technology revolution is not a one-off conversion. It has, like the steam engine, raised the pace of technical change for a generation, creating a need for more flexible small-batch production systems, a better-educated more innovative workforce, and more creative, individualistic management. Growth will come in new products, knowledge-intensive products, more differentiated than mass-produced, with more of their value in design ... In high-technology industries, whoever pioneers production will have a decisive influence on future patterns of trade, with others finding it hard to catch up.

While technological change cannot and must not be considered in isolation from equally radical changes in company organisation, labour practices and skills, and market demand, its role does therefore seem to be of fundamental importance in the industrial restructuring which is currently affecting firms, industries and local environments in the United Kingdom.

NEW TECHNOLOGIES AND EXISTING INDUSTRIES

Research into technological innovation has conventionally distinguished between "process" and "product" innovations, the former involving adoption of a significantly new method or process of manufacturing existing products, the latter the development of an entirely new product itself. As Breheny and McQuaid (1987, 302) point out, the phrase "high-technology industry" is almost always taken to refer to cases of the latter, where new technologically-advanced products developed from considerable inputs of scientific research are manufactured by new industries. However, it is clear that the impact of new process technologies - themselves often the outputs of high-technology producer industries - upon production in **existing** industries is also of great significance for localities and regions. Some brief reference to this impact in the UK case is therefore necessary here.

Firstly, it seems very probable that investment in new technologies by existing industries is one major reason for the high rates of growth of manufacturing output per head achieved since 1981 (Figure 4.1). Such investment has been a key feature of company strategy, along with largescale labour redundancies and a shift to new flexible work practices, in many large firms during the 1980s (Peck and Townsend, 1987). This seems to be especially true of foreign-owned corporations, both longstanding and new to Britain. A striking example is Pirelli UK, the West Midlands-based tyre manufacturer, where massive job losses in the early 1980s were followed by a combination of adoption of new process technologies and radically new work practices. The resultant dramatic increase in labour productivity moved the company's main Burton-on-Trent plant from 117th to 5th in this respect in the world-wide league table of Pirelli plants 1981-6, with its Carlisle plant reaching top position. Other recent management innovations, including Just-in-Time methods, flexible work teams, and much closer sales force/production integration, are being coupled with a 1987 £35-million investment in new technology designed to raise labour productivity by a further 50% (Leadbeater,

1987a).

Another example of new technology investment in a traditional industry, this time by a very new foreign multinational, is Nissan UK. The company's first-ever European car assembly plant, opened at Washington, Tyne and Wear, in 1986, utilizes advanced technologies to manufacture 29 thousand cars a year with only 470 workers. The plant's efficiency and labour productivity is believed to be equalled by only one other assembly plant in Europe, and the decision to expand it to 100 thousand-unit capacity and 2,700 jobs by 1991 was brought forward by two years partly because of its initial success. This technologically innovative development in an old industrial region of high unemployment is also now attracting Japanese component firms to begin production in this area.

Of course, larger British-owned companies in traditional industries have also been investing in new technologies, in some cases to a marked degree. The UK textile industry, despite enormous job losses and mill closures in the 1970s and early 1980s, has seen considerable investment in computer-controlled spinning and weaving by large companies such as Vantona Viyella. Williams (1985) reports the latter's claim that "new technology has reduced production costs so much that in many areas we can compete with low-cost developing countries". He also notes the National Wool Textile Export Corporation's view that "for spinning, the UK is probably the most technically advanced country in the world", with sharp increases in labour productivity since 1983. Employment in the UK textile industry as a whole nonetheless has continued to decline, by 56 thousand or 20%, 1981-4.

Other UK industries which have invested heavily in new process technologies in the 1980s include food processing and oil refining/petrochemicals (Butchart, 1987, 84). The former illustrates particularly clearly the radical nature of "flexibilisation", involving both new technologies and changing labour practices. In the case of KP Foods, a subsidiary of United Biscuits, 85% of its 6,000-strong, mainly female, production workforce is now part-time, compared with a majority of full-time workers in 1980. This radical change, effected at both the company's Ashby-de-la-Zouche (Leicestershire) and Grimsby factories, was necessary, according to the company, in order to raise productivity and increase production flexibility in the context of a £60-million investment programme in new technology, involving computer-controlled mixing, packaging and flow-line production. The latter was in turn a direct response to changing market demand, towards increasingly short-life "customized" products marketed by retailers under "own-brand" labels (Leadbeater, 1987b).

The local and regional impact of restructuring with new technologies in existing industries is not easy to assess. But the above examples and recent work by Peck and Townsend (1987) suggest that in the UK case, it has mainly affected large companies and plants in the country's traditional 19th-century industrial regions, such as the West Midlands and North West England, and is associated with substantial job losses there. In most cases, this has been coupled with an increasing focus on more-skilled male, rather than less-skilled female, workers. New technology implementation also not infrequently involves some shift of production out of congested conurbation areas, as well as to assisted regions, where government financial incentives for new buildings and production equipment can be accessed. The latter is of course illustrated by the Nissan case discussed above.

THE NATURE AND DEFINITION OF HIGH-TECHNOLOGY INDUSTRY

As noted in Chapter 1, different underlying assessments and perspectives on current new technologies suggest different definitions of high-technology industry. The approach adopted here follows the second of the two alternatives identified, whereby all radically new technologies which have appeared in recent years, and not just those in the information technology field, are viewed as potential contributors to the development of a new long cycle of economic growth. It also therefore logically focusses on the impact of significant innovations in creating new products and industries, and the essential role of research and development personnel and expenditure in this process.

This approach to definition has led to the identification of sets of high-technology manufacturing industries such as that proposed by Kelly (1986, 59-73) on the basis of R and D activity and product innovation rate, and used by Keeble (1987, 16). One problem here, however, is that the manufacturing sector as conventionally defined excludes certain highly technologically innovative **service** activities, such as separate R and D laboratories and computer software establishments. Another problem is the rapid rate of technological change itself, such that high-technology definition is always addressing a moving target. On both grounds, therefore, it would seem appropriate to adopt for use in this chapter the very recent definition proposed by Butchart (1987) on the basis of work in the UK Department of Trade and Industry. This includes for the first time three service activities along with 16 manufacturing sectors. The two criteria employed by Butchart are above-average R and D intensity (ratio of R and D expenditure to value of gross output), and proportion of scientists, professional engineers and technicians in the workforce. These two measures are precisely the key industry-level indicators of high-technology status focussed on by virtually all recent empirical studies, such as those by De Jong in the Netherlands (1987, 37-40), Thompson in the USA (1987), and Breheny and McQuaid in the UK (1987). The resultant list of industries is given in table 4.1. The usual qualifications concerning unavoidable internal heterogeneity of SIC sectors with respect to technological intensity, and variations in technological sophistication and function between establishments in the same SIC category but located in different areas, of course apply.

Defined in this way, high-technology industry in the United Kingdom possesses a number of characteristics of relevance to understanding its local and regional evolution and impact. Firstly, unlike the situation in the USA, employment in UK high-technology industry has fallen, not risen, in the last decade and during the 1980s. The total for Great Britain rose from 1.305 million in 1976 to 1.342 million in 1979, only to fall thereafter, to 1.239 million by 1986 (Butchart, 1987, 88). The latter is however a much slower rate of loss (-7.7%) than for manufacturing as a whole (-27.7%), so the _share_ of high-technology industry in manufacturing employment is growing. Equally, high-technology manufacturing industries are clearly high-growth industries in _output_ terms, the volume of production in these industries growing by +46% 1976-85 compared with a _decline_ in total manufacturing output of -2% (Butchart, 1987, 87). At the aggregate level, then, high-technology industries would seem to offer significantly better prospects of job stability and economic buoyancy for local and regional economies than conventional manufacturing industries, notwithstanding some national employment decline. This is fully borne out by the actual experience of particular localities such as

Table 4.1. High-Technology Industries in the United Kingdom

SIC	Industry Description	SIC	Industry Description
2514	Synthetic resins & plastics materials	3453	Active components & electronic sub-assemblies
2515	Synthetic rubber	3640	Aerospace equipment manufacturing & repairing
2570	Pharmaceutical products	3710	Measuring, checking & precision instruments & apparatus
3301	Office machinery	3720	Medical & surgical equipment & orthopaedic appliances
3302	Electronic data processing equipment	3732	Optical precision instruments
3420	Basic electrical equipment	3733	Photographic & cinematographic equipment
3441	Telegraph & telephone apparatus & equipment	7902	Telecommunications
3442	Electrical instruments & control systems	8394	Computing services
3443	Radio & electronic capital goods	9400	Research & development
3444	Components other than active components mainly for electronic equipment		

Source: Butchart, 1987

Berkshire (Hall et al., 1987, 184) and Cambridge (Keeble, 1988).

A second significant characteristic of many if not all high-technology industries concerns the changing balance of large and small firms. There is no doubt that many of these industries, such as aerospace and telecommunications, are dominated by large, usually multinational, companies. Even in the computer electronics industry, only four firms - IBM, ICL, Digital Equipment and Honeywell - accounted for no less than 70% of all UK computer sales in 1985 (Kelly, 1987, 15-17). The performance, strategies and investment decisions of such large companies are thus of crucial importance for local economies. However, this said, it is also true that in a number of high-technology sectors, the last ten years have witnessed a substantial growth in the numbers and importance of small, usually new, firms and establishments. This shift in organisational structure is illustrated for the computer industry case (hardware and software) by table 4.2.

Table 4.2 Changing Firm Size Structure in the UK Computer Industry

Establishment Size (Emps)	Employment in 1980	(%)	Change 1980-4
0 - 50	3,467	(8.8%)	+11,268
51 - 100	1,644	(4.2%)	+1,193
101 - 250	2,697	(6.8%)	+982
251 - 500	2,786	(7.1%)	+915
501 - 1,000	2,036	(5.2%)	+1,083
1,000 plus	26,765	(67.9%)	-4,044
All estabs.	39,395	(100.0%)	+11,397

Source: Kelly and Keeble, forthcoming

The growth in employment in firms and establishments which were entirely new or employed less than 50 workers in 1980 was particularly impressive. Since 1975, relatively large numbers of new, small independent companies have been set up in this industry, usually by highly-qualified entrepreneurs exploiting new market opportunities created by very rapid technological change (Keeble and Kelly, 1986). This process, which underpins recent high-technology growth in such areas as Cambridge, also characterizes other high-technology sectors such as research and development services, medical equipment and scientific instruments. Kelly (1987, 118) links new firm formation in such sectors to the flexibility they afford "boffin" entrepreneurs in responding to very rapid technological change, as well as a proliferation of specialised market niches suitable for exploitation by small firms. While many of the more successful of such firms may well be targets for large-firm acquisition at a later stage, as noted by Pottier (chapter 5), their creation and growth is both rooted in and benefits specific, often previously less-industrialised, local environments. It should be noted that new small firms are frequently engaged in R and D or specialised manufacturing, mass production of high-technology products being generally the province of large corporations.

The last structural characteristic of UK high-technology industry which will be emphasised here is its fundamental dependence upon highly-qualified staff, in the form of research scientists, engineers, technicians, managers and professionally-qualified workers. Many researchers agree that the single most vital input in most successful high-technology companies is **brain-power:** continuing and effective scientific research is essential for competitive success in a technologically-dynamic environment. The resultant growth in dependence of many UK high-technology industries upon highly-qualified staff during the 1980s is graphically illustrated by computer manufacturing. In this industry, the number of R and D workers nearly **doubled** (+93%) between 1978 and 1984, whereas its low-skilled production workforce **declined** by 20% (Kelly and Keeble, forthcoming). This rapid shift in occupational structure towards highly-qualified personnel is common to all information technology industries, and shortages of such staff are now quite considerable. Their significance is equally great both for large firms, with large R and D laboratories and programmes, and small firms, most of which are set up by technically-qualified individuals acting as entrepreneurs. Thus Kelly's work (1987, 137) shows that 86% of all new computer firm founders in the UK possess degree-level or equivalent technical qualifications, with 54% possessing Ph.Ds or computing/electronics degrees.

HIGH-TECHNOLOGY INDUSTRY: THE SPATIAL PATTERN

The broad regional distribution of UK employment in high-technology industry, as defined by the new Butchart classification, is recorded in table 4.3. Employment figures are the only detailed measure of the spatial distribution of such industry currently available, and are derived from Census of Employment records for 1981 and 1984. Three features of these hitherto unpublished statistics deserve emphasis. First, high-technology activities are clearly heavily concentrated in southern regions, and especially South East England. The latter region alone contained no less than 44% of total British high-technology employment in 1984, with 56% or nearly three-fifths in the three southern regions of the South East, South West and East Anglia.

Table 4.3 The Regional Distribution of High-Technology Industry in Britain, 1981-4

	1981 Employees	1984 Employees	Change 81-4 No.	%
East Anglia	30,802	37,231	+6,429	+20.9
Wales	36,926	37,993	+1,067	+2.9
Rest of South East	321,085	322,127	+1,042	+0.3
Scotland	87,695	86,044	-1,651	-1.9
North	46,286	43,864	-2,422	-5.2
Greater London	211,492	200,058	-11,434	-5.4
East Midlands	78,013	73,391	-4,622	-5.9
West Midlands	104,123	97,469	-6,654	-6.4
South West	114,744	105,554	-9,190	-8.0
Yorkshire and Humberside	52,328	46,540	-5,788	-11.1
North West	145,072	127,870	-17,202	-11.9
Great Britain	1,228,566	1,178,141	-50,425	-4.1

Source: Unpublished Department of Employment statistics

Secondly, the regional evolution of high-technology activity during the 1980s reveals that only three regions experienced overall net growth of high-technology jobs, these being East Anglia, Wales and outer South East England. The three next best performing regions, with only relatively small rates of job decline - and almost certainly above-average output growth - were Scotland, Northern England and Greater London. This pattern suggests that two different types of region have been attracting or generating new high-technology investment in the 1980s. One comprises a broad swathe of south-eastern England, and particularly its less-urbanised outer areas such as East Anglia. The other is the high-unemployment assisted industrial regions, notably Wales and Scotland. The very rapid growth of East Anglia (+6.4 thousand jobs or 21% in only three years) is particularly striking.

Thirdly, the old industrial regions of North West England and Yorkshire and Humberside suffered the greatest losses of existing high-technology jobs, with the South West also suffering above-average decline. The latter, and a significant part of the North West's losses, are however almost certainly to be explained by the bias in the structure of their high-technology activity towards **aerospace manufacturing**, which has witnessed by far the greatest employment decline amongst high-technology sectors in the 1980s, through restructuring and job shedding by the industry's large corporations (for example British Aerospace, with major production centres at Bristol and in Lancashire).

A much more detailed picture of the contemporary geography of high-technology activity in Britain is presented in Figures 4.2 to 4.4, based on unpublished Census of Employment data for the counties of England and Wales and regions of Scotland. Figure 4.2 confirms the contemporary importance of Greater London and other South East counties - Hertfordshire, Berkshire, Surrey and Hampshire - for high-technology activity, with secondary concentrations in North West England and central Scotland. Recent trends in the spatial evolution - growth and decline - of high-technology industry between 1981 and 1984, as shown for the first time in Figures 4.3 and 4.4, are perhaps more interesting.

The first of these reveals that not surprisingly in view of the aggregate national trend noted earlier, a majority of counties recorded net losses of high-technology jobs during the 1981-4 period. The biggest declines were experienced

Figure 4.2 The Geographical Distribution of High-Technology Industry in Great Britain, 1984

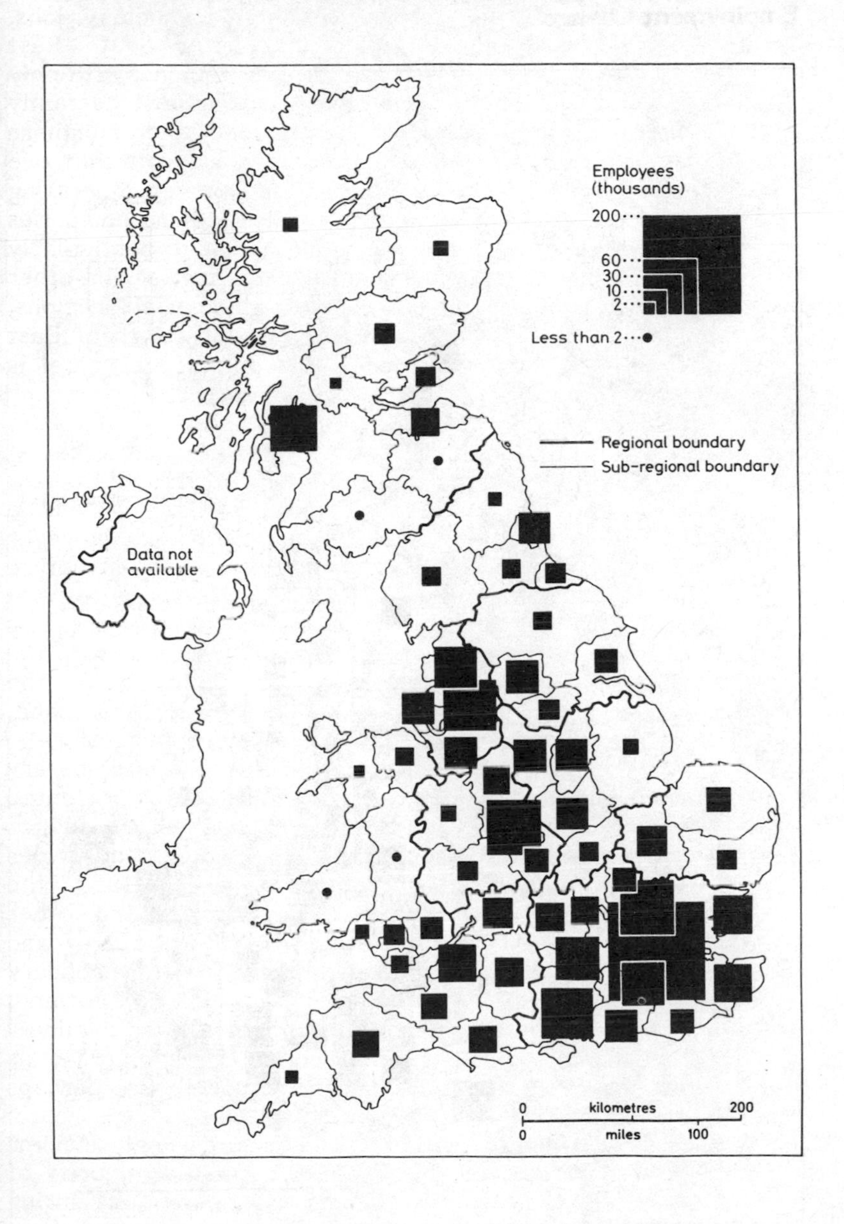

Figure 4.3 The Geographical Evolution of High-Technology Industry in Britain, 1981–4, by Volume of Employment Change

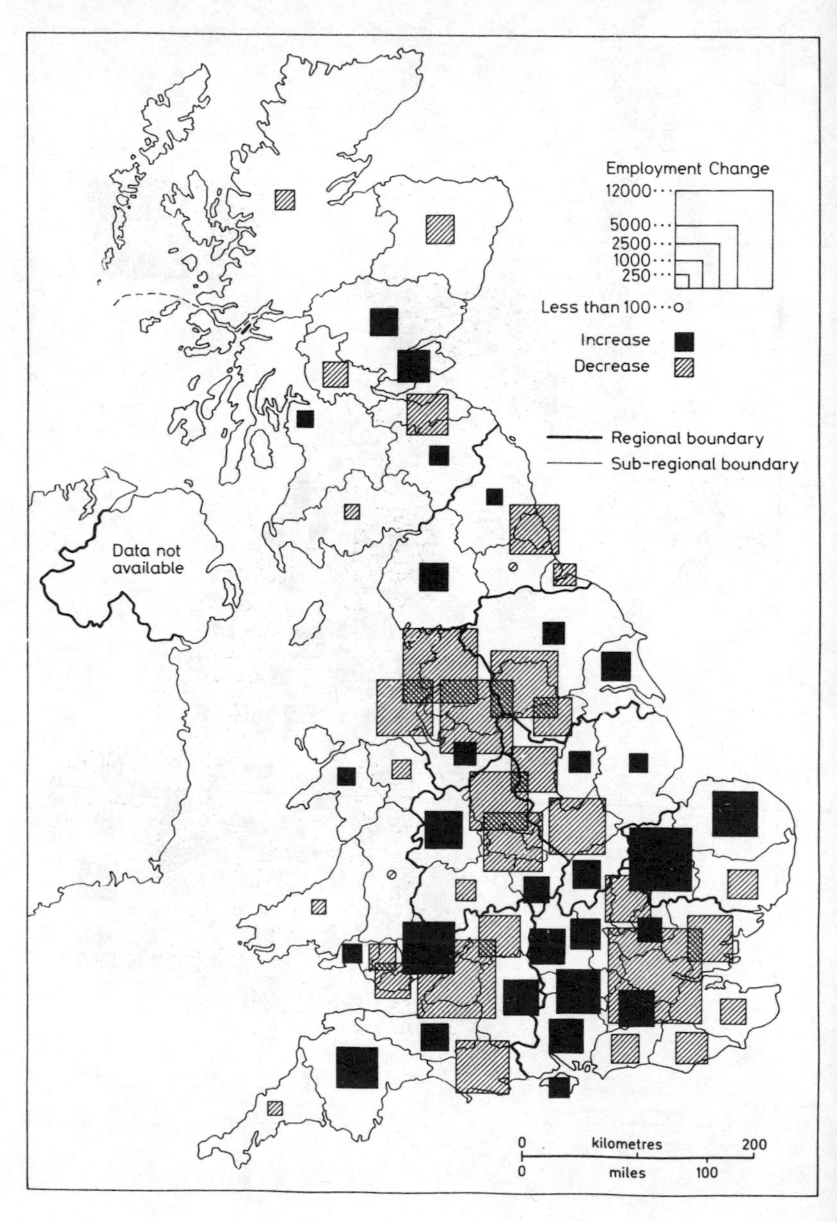

Figure 4.4 The Geographical Evolution of High-Technology Industry in Britain, 1981-4, by Rate of Employment Change

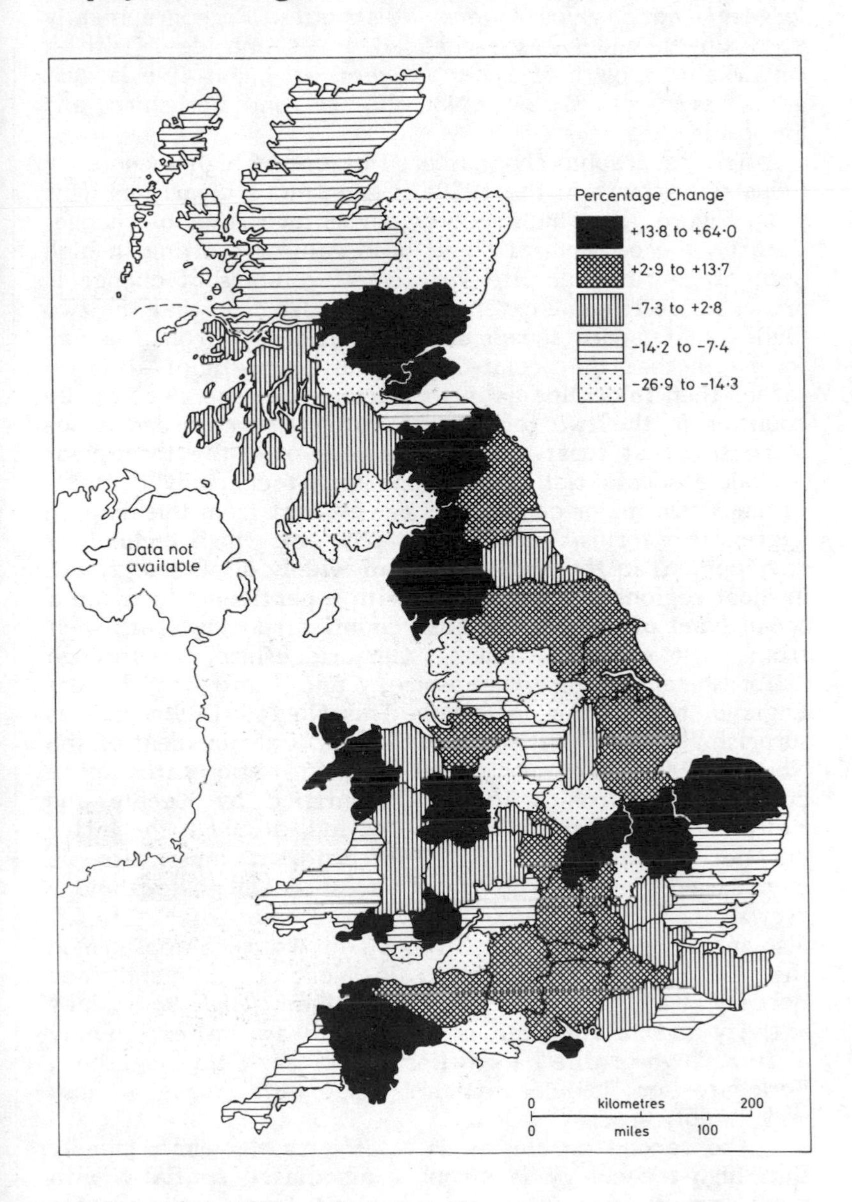

by London, Avon (Bristol), Manchester, Lancashire and other major urban centres such as West Yorkshire (Leeds), West Midlands (Birmingham) and Merseyside (Liverpool). The largest net gains were distributed geographically surprisingly widely, spearheaded by Cambridge, with a remarkable growth of 5,000 workers. The next five largest gains were by Gwent, Norfolk, Devon, Berkshire and Shropshire.

The geographically-scattered nature of high-technology industrial growth in the 1980s is even more clearly evident from Figure 4.4, which records percentage rates of change. Clearly, the problem of small base values resulting in high percentage rates despite very small volumes of change is present here to some extent. But it is also true that the two highest categories shown differ significantly from the two lowest in that they relate to absolute **growth** of activity, rather than to **decline** as in the latter case: and 14 of the 26 counties in the two top growth categories recorded a net increase of at least 1,000 extra jobs over the three-year period, a substantial total in the high-technology context. Perhaps two major conclusions are evident from this map. In aggregate terms, **the geography of high-technology development in the 1980s is one of widely dispersed growth in most regions of Britain, but with a particular focus on a broad band of southern English counties** running south-west from Norfolk and Cambridgeshire, through Oxfordshire/Berkshire/Hampshire, and into south-west England, terminating in Devon. This Norfolk-Devon axis is surprisingly - and perhaps significantly - reminiscent of the Norwich-Bristol axis of high new firm formation rates in the computer electronics industry identified by Keeble and Kelly (1986, 86) for the 1975-84 period, although the latter did not extend to Devon. That said, it must also be emphasised how geographically widespread high-technology growth is, with the fastest-growing category of Figure 4.4 also including Gwent and Gwynned in Wales, Shropshire in the West Midlands, Cumbria in Northern England, and Borders, Tayside and Fife in Scotland. High-technology activity and development is thus in no way confined to only a few "high-profile" localities, such as Cambridgeshire, Berkshire or "Silicon Glen", important though these undoubtedly are.

The second conclusion is related to the first, namely that **high-technology is strongly associated spatially with rural, small town and less-urbanised local environments**

throughout Britain, whereas large cities and the conurbations are generally characterised by high-technology **decline.** This is strikingly illustrated by table 4.4, which records the distribution of the top and bottom 20 counties by rate of high-technology employment growth (above +8.6%) and decline (below -10.8%), classified by their degree of urbanisation. The urbanisation definitions are taken from Keeble (1980, 962) and are based on the presence or absence in a particular county of towns or cities of a particular size, plus the county's overall population density.

Table 4.4 Urban-Rural Contrasts in High-Technology Employment Growth and Decline, 1981-4

Urbanization Status	High-Technology Employment Change Most Rapid Growth*	Most Rapid Decline**	All GB Counties
Rural counties	10	5	20
Less-urbanized counties	8	2	21
More-urbanized counties	2	8	14
Conurbations	0	5	8

* above +8.6%   ** below -10.8%

As the table shows, all but two of the 20 "highest-growth" counties are rural or less-urbanised areas, characterised by only small towns (generally less than 75,000 inhabitants) or villages: whereas 13 of the 20 "highest-decline" counties are conurbations or more-urbanised areas. Indeed, as the table shows, three-fifths of **all** conurbations and more-urbanised counties record rates of high-technology decline greater than 10.8%. The only exception to the clear association between high-technology industry growth or decline and urban status is the group of five rural counties in the highest decline category; three of these do however involve only small absolute volumes of loss, though high rates. The reasons for the generally striking association between rural and small town environments and high-technology industry growth will be considered in the next section.

THE DETERMINANTS OF HIGH-TECHNOLOGY INDUSTRY LOCATION

The current location of high-technology activity within the UK and other European countries would not seem to be easily or adequately explicable in terms of traditional or abstract location theories, whether of the Weberian or Marxian kind. This is because the key ingredients and conceptual underpinnings of these theories, such as transport cost minimization or cheap labour exploitation, are either irrelevant to or only very partial factors in initiation, location choice and competitive success for most high-technology companies, at least at the intra-national scale. Explanation would seem rather to lie in the analysis of how macro-economic trends of technological change and market demand have impacted upon and interacted with inherited local social and economic structures, in complex and often place-specific processes of historic evolution. While key macro-economic forces can be identified underpinning the locational evolution of high-technology activities, the role of unique local environments, of locally-specific inherited industrial, occupational and even physical characteristics, is also of great importance in this respect. This section thus attempts to isolate key forces whose importance for competitive success have channelled high-technology industry to particular kinds of areas, while the subsequent section highlights the vital role of particular local environments and the different types of high-technology development which have evolved out of different local socio-economic structures.

For most high-technology firms in Britain, it can be argued that the single most influential determinant of their locational evolution in the 1980s is **the spatial distribution of highly qualified manpower**, in the form of research scientists, engineers, managers, technicians, and professionally qualified workers. This contention is supported by both general and specific arguments. For activities which by definition are operating in a rapidly-changing and highly-turbulent technological environment, competitive success is in general crucially dependent on being ahead of the field in terms of R and D and technological leadership. The extraordinary decrease in product life cycles - from 10-15 to only three or five years - witnessed in recent years in many high-technology sectors in the context of intense global competition directly necessitates an ever-increasing dependence upon R and D and research staff. The very rapid recent growth in numbers and proportion of such staff in the UK computer industry noted earlier illustrates this clearly.

In specific terms, also, as noted earlier, most new high-technology firms are actually and directly set up by highly-qualified entrepreneurs, while large companies are expanding their research staff and R and D laboratory activity rapidly relative to production workers. A good example of the latter is Digital Equipment Corporation, whose 1,000-job (+18%) UK expansion programme announced in 1987 is heavily focussed on its Reading R and D laboratory (350 extra staff, to a total of 950), and recruitment of computer engineers for locations in south-east England (over 60% of the 1,000 total). Another case is McDonnell Douglas Information Systems, whose British information technology division is also being expanded rapidly with a projected 700-job growth 1987-9, the largest single component of which will be 300 additional research and support personnel to be recruited to its Hemel Hempstead, Hertfordshire, R and D centre.

The key point here, however, is of course that highly-qualified personnel are not only in short supply, but highly-selective and "choosy" with regard to the local environment in which they - and their families - are prepared to live and work. A number of empirical surveys show that such individuals have clear, and highly-differentiated, **residential space preferences,** focussed most particularly on small town and rural environments in southern England. The latter are precisely the kind of localities to which such individuals have been migrating, in large measure for environmental reasons (Keeble and Gould, 1985: Keeble, 1986, 179-82: Perry, Dean and Brown, 1986, 92-4), as a component of "counter-urbanisation" migration over the last 15 to 20 years. The marked bias in current high-technology industry growth to such areas, both in the form of new firm creation by previously or concurrently migrant founders (Keeble and Kelly, 1986: Keeble and Gould, 1985) and core R and D and prototype development by large companies, can thus directly be related to the sector's vital dependence upon highly-qualified staff and entrepreneurs. These are most readily present or recruitable in less-urbanised areas of high perceived residential amenity, such as Cambridge (Keeble, 1988), Berkshire (MacGregor, Langridge, Adley and Chapman, 1986, 441), or Devon and Cornwall (Spooner, 1972, 208).

This key macro-economic factor overlaps with another; namely, the existence in particular areas, often for historic reasons, of a visible and substantial **local scientific research capacity** in the form of major scientific universities, public sector research establishments, or large firm research laboratories. Such existing research concentrations can provide a local pool of experienced researchers, a steady flow of new graduate recruits for high-technology companies, a supply of entrepreneurs through spin-off from existing R and D activity, and even a source of technological innovations through informal and formal research linkages. Thus the special role of Cambridge university and its major scientific research capacity is central to the "Cambridge phenomenon" (Keeble, 1988), while Hall, Breheny, McQuaid and Hart (1987, ch. 9) argue that the location of high-technology industry in Berkshire has been strongly influenced by the opportunities for informal linkages with the government defence research establishments which are concentrated there.

At least three important qualifications concerning this important location determinant must however be stressed. First, it is clear that the prior existence of local research capacity is not in itself a sufficient - nor, indeed, may even be a necessary - condition for successful high-technology development, even of an R and D-focussed kind. The "artificial" development of major R and D-based high-technology activity in the Côte d'Azur of France since the 1960s by a process of deliberate colonization by multi-national firms of a hitherto unindustrialised but residentially very attractive "sunbelt" location is a good example here. So too is the difficulty faced by attempts in Britain's old industrial conurbations, as by Salford University in Manchester, to promote local high-technology development despite deliberate technology transfer policies. Second, the stimulus of existing research capacity is largely confined to the research and prototype development end of the high-technology spectrum: the location of mass-production of high-technology products is chiefly influenced by other considerations (see below). Third, the historic development of existing research capacity is often the product of other, and perhaps therefore more fundamental, influences, such as the role of residential preference in the Côte d'Azur case, or the great importance of market proximity in the 1930s evolution of London's radio and electrical engineering industries, from which the "western crescent's" R and D-based information technology industries have developed and diffused.

For research-focussed activity, the last general locational influence which warrants emphasis is **proximity, via motorway and high-speed rail communications, to London and its international airports.** The importance of such proximity for multinational high-technology companies is clear and attested by many studies (e.g. Hall, Breheny, McQuaid and Hart, 1987, ch. 11): but it is also very important for smaller independent firms, given their unusual degree of national and international marketing activity compared with small firms in traditional sectors. Thus Kelly (1986, 279) shows that 55% of all 1984 sales by new computer firms in the UK were to wider national (27%) or international export (28%) markets, beyond the region in which the firm had been founded. Proximity to London's financial markets for investment capital is also probably an advantage to small high-technology firms in southern England.

The above discussion has focussed on the vital importance of R and D and prototype development in high-technology success and location. However, large companies are of course able to separate R and D and mass-production activities, and to gain economies of scale and take advantage of different environmental requirements by locating these in different areas (Sayer and Morgan, 1986). The importance of the resultant **functional and spatial division of labour** in high-technology activity in the UK must not be underestimated. As Kelly's work shows (1987: Kelly and Keeble, forthcoming), this division is longstanding, involves a significant difference in occupational structure and quality of high-technology jobs as between southern and peripheral Britain, and is primarily due to the inward investment activities of American and other foreign multinational corporations. Their establishment since 1945 in assisted regions such as Scotland (Silicon Glen) and Wales of mass-production branch plants manufacturing semi-conductors, computing and data-processing equipment, and pharmaceuticals has created in these areas an entirely new, "implanted", set of high-technology activities. Why has this occurred, and what macro-economic influences are involved?

The first point to stress is that this phenomenon clearly reflects the utility of a British base for American and Japanese firms in particular as a production platform for serving the wider European - and especially European Community - market. As Kelly (1986, 213) points out, the **European** orientation of Scottish high-technology production is strikingly illustrated by the fact that eight of the top ten US/Japanese information technology companies now operate a unit there, whereas this is true of only one (Ferranti) of the top ten European IT companies. Within this global context, however, detailed assessment of the precise history, nature and process of high-technology branch plant establishment in Scotland, Wales and Northern England leaves little doubt that the single most influential reason for the choice of such locations within Britain has been the impact of **government regional policy**. This has operated at particular periods with great force in terms of factory controls, financial grants and subsidies, and persuasive and effective promotional organisations such as the Scottish and Welsh Development Agencies. As a result, 80 per cent of the foreign-owned computer electronics establishments surveyed by Kelly (1986, 214) in Scotland reported that regional policy incentives had been of major importance in their decision to locate there.

The continuing current flow of foreign high-technology investment to the assisted regions of Britain under the stimulus of regional policy incentives and agencies is also well illustrated by the finding of the 1987 US Electronics Survey (Townsend, 1987) that not only is "Britain by far the preferred location of new US high-technology investment" planned in Europe over the next two years, but that within Britain, it is the assisted regions which are most highly favoured, with Scotland in particular projected to gain more new US high-technology jobs (3,700) 1987-9 than any other region of Europe. This reflects a regional incentive package promoted particularly effectively by the Scottish Development Agency, whose success, like that of the Irish Industrial Authority, "reflects early realization of the electronic industry's potential, followed by consistent efforts to attract investors. Both have created an environment in which electronics companies can flourish and have won financial backing from the government" (Townsend, 1987).

This crucial policy influence on assisted region mass-production activity, reflecting _inter alia_ its sensitivity to capital incentives because of the expense and scale of new process technology required in such plants, has been reinforced by certain secondary **labour market advantages.** Female production workers and locally-trained technologists and engineers are relatively readily available in regions such as Scotland, as a consequence of high regional unemployment and a tradition of technological education (Henderson and Scott, 1987, 68-71). It should be noted however that operative wage levels in these regions are surprisingly similar to those elsewhere in Britain, while labour militancy, as measured by the incidence of industrial disputes, is and has been for many decades appreciably higher. Moreover, as Sayer and Morgan (1986, 160) point out, the largest single **absolute** concentration of production activity and less-skilled operatives in most high-technology industries, such as electronic capital goods, is actually to be found in South East England, not in the assisted regions, because of the dominance of this area in high-technology activity generally. These considerations support the argument that labour market advantages for mass production are secondary rather than primary influences on the evolution of high-technology industry in peripheral Britain, as part of a relative functional and spatial division of labour in this sector.

LOCAL ENVIRONMENTS AND THE EVOLUTION OF HIGH-TECHNOLOGY ACTIVITY

The preceding section has sought to isolate and highlight the key influences which appear to underlie and explain the geography of high-technology industry in the UK in the 1980s. The latter is however undoubtedly in a state of rapid change and evolution, perhaps not surprisingly in view of the global and technological dynamism of this sector. The Cambridge phenomenon, for example, has only been clearly evident since 1981 (see above), the county recording net **losses** of high-technology jobs before this (Breheny and McQuaid, 1987, 316). The latter was almost certainly due solely to large plant restructuring by Cambridge's only existing multinational, Philips (originally Pye of Cambridge). Again, Oxfordshire's significance for advanced technology industries, with a 1986 employment total nearly three-quarters that of the Cambridge region, has only very recently been recognised and charted for the first time by Lawton-Smith (1987).

More important still, however, the macro-level

determinants identified above have acted in different and complex ways in different local historical, social and economic environments. The working out of these macro-level forces "on the ground", in particular localities, has been a complex and highly diversified process, in which **the precise mix of local environmental conditions** has been of great significance for the extent and nature of high-technology development. This is true both for localities, such as Cambridge and Oxfordshire, where high-technology development has been largely an indigenous, spontaneous and entrepreneurial process, and those, such as south Wales and central Scotland, where it has been implanted from outside by multinational branch plant colonization of an old industrial region. The final part of this chapter will thus seek to illustrate the key role of local environments, of inherited locally-specific social and economic structures, in influencing and shaping the nature of high-technology activity in three arguably very different high-technology locations, namely Cambridgeshire, Berkshire and central Scotland.

The Cambridge phenomenon

Until the 1970s, high-technology industry in the Cambridge region was dominated by a few historic companies such as Pye radio, TV and telecommunications, taken over by Philips in 1967 and subsequently rationalised, with a loss of 1,800 jobs or 35% by 1984. Since the early 1970s, however, a remarkable rate of new company formation has resulted in a current total of 350 high-technology firms, employing 16,500 workers or 11.4% of total employment (manufacturing and services) in southern Cambridgeshire. Three-quarters of these have been set up as entirely new, independent companies by individual entrepreneurs, in activities ranging from electronics and computers to scientific instruments, biotechnology, pharmaceuticals and R and D consultancy. Most companies are small, and tend to be R and D rather than production focussed; but a significant number have grown rapidly, while incoming multinational companies such as Schlumberger, Napp Pharmaceuticals, GEC-Marconi and London International are now adding to the high-technology jobs boom clearly evident in Figure 4.3. Most of this activity is research-focussed, with no less than 47% of jobs in scientific and technical occupations and a further 10% in managerial occupations: only 22% are engaged in production-focussed manual jobs, skilled and unskilled.

Recent work by the present author (Keeble 1987b) argues that the key to the development of this so-far highly successful "technology-oriented complex" lies in the historical evolution in and attraction to the area of a pool of exceptionally talented research scientists, engineers and technologists. This pool has provided the "boffin" entrepreneurs, highly-skilled staff and, above all, the technological innovations on which competitive success, often in an international marketplace, has been based. The existence of this pool reflects two fundamental environmental features of the area. One is the role of the university, as an historic and globally-important centre of scientific and technological research and university education, while the other is the unique environmental attractiveness of the city as a place to live and work, because of its historic architecture and cultural vitality. The traditional absence of largescale manufacturing industry, however, has inhibited the development of a supply of less-skilled production workers, while their recruitment from outside the area is now heavily constrained by housing shortages and high housing prices. This distinctive socio-occupational characteristic of the local labour market is of key importance in explaining the bias in high-technology development towards research-based "soft" activities (Segal Quince and Partners, 1985, ch. 6) rather than "hard" manufacturing production. The latter has often been sub-contracted by Cambridge firms to older industrial regions such as Scotland (Sinclair) and Wales (Torch).

Other aspects of the local environment and Cambridge's historic economic evolution which have helped shape current high-technology development include particular policy decisions of central government, notably the establishment in Cambridge in 1969 of a computer-aided design centre from which many new firms have spun off, the active intervention of a leading national bank to promote local development, Cambridge's proximity to, and rapidly-improving communications (the M11 motorway and electrified rail links) with, London, and a generally supportive political and institutional environment in terms of local planning and university policies. The latter is most strikingly illustrated by the pioneering development by Trinity College since 1970 of the Cambridge Science Park, the most visible symbol of Cambridge's high-technology success.

Berkshire and the Western Crescent

As previously demonstrated (Figure 4.2), the largest single concentration of high-technology activity in Britain outside London is to be found in a broad arc or crescent running round the western edge of London from Hertfordshire to Hampshire. This "western crescent" as Hall et al. (1987, 47) have termed it is centred on the county of Berkshire, which in the 1975-81 period recorded the largest volume (+7,600 jobs) of high-technology employment growth in the country. Berkshire's expansion has continued in the 1980s, with the fourth largest volume of new high-technology jobs in Britain 1981-4 (Figure 4.3). However, the nature of high-technology development here, and of the local milieu which has engendered and shaped it, differs in important ways from the Cambridge case. Thus Berkshire contains a range of large, medium and small companies, while the scale of activity is much greater, and involves considerable production as well as major R and D facilities. And in sharp contrast to Cambridge, local university research has played virtually no part in high-technology development, Berkshire's scientific research base being provided by government research establishments (GREs) and private-sector British and multinational company research laboratories.

Environmentally, the "Royal County of Berkshire" ranks highly as a residentially-attractive area of smaller towns (Windsor, Maidenhead, Newbury) and villages set in pleasant wooded landscapes on either side of the river Thames. Although the county's largest town, Reading, possessed some industry before 1939, economic and demographic expansion really came after World War II, epitomised by the designation of Bracknell as one of London's eight postwar government "new towns", to take decentralised manufacturing and population from the congested capital. Industrial decentralisation from west London to Berkshire as a whole developed rapidly under the stimulus of industrial expansion and space constraints in the capital (Keeble, 1968). The county's position athwart the main Bristol-London railway line and the construction of the M4 motorway, along with its attractive residential environment, also encouraged a rapid growth of commuting population, heavily biased towards managerial and professional workers.

The high-technology growth of the last 15 or so years has thus occurred in the context of an already fairly substantial local industrial sector and labourforce, along with excellent communications, including the close proximity and accessibility of London Heathrow, the world's busiest international airport. The scale of high-technology industry is now considerable, with a county total (1984) of 37 thousand jobs, double that of Cambridgeshire, in a wide range of sectors such as defence electronics, electronic components, computers and data processing, and telecommunications. Processes of high-technology growth have involved the continuing expansion of larger firms which originally moved out from London (Racal at Bracknell, for example), the arrival of multinational American companies which have established R and D and European-servicing facilities, and spin-off of many small entrepreneurial companies from existing firms.

Decentralised companies were often already operating in a high-technology sector because west London, the area in which most of them originated, was itself Britain's leading centre of technologically-advanced industry (radio and TV, electrical engineering) in the 1930s. In turn, however, this historically rooted spatial diffusion process of high-technology firm relocation has helped create a pool of highly-qualified R and D staff which, along with Berkshire's exceptional international and national communications nodality and proximity to London, has been a powerful influence attracting foreign multinationals. For both types of firm, the area's residential attractiveness to highly-qualified staff is a very important locational advantage, while this also underpins a high rate of new firm formation by high-technology entrepreneurs. The latter has also been encouraged by local market and sub-contracting opportunities, given the rapid growth in volume of local high-technology production since 1970.

A very important question concerning the evolution of Berkshire's high-technology complex which has not so far been discussed is the role of defence procurement and government research establishments (GREs). In their recent extensive work, Hall et al. (1987, chs. 8 and 9) argue that the location and growth of numerous defence-focussed GREs in the western crescent, including Berkshire, together with substantial growth of government purchasing of sophisticated defence equipment since 1950, has provided a powerful and indeed dominant stimulus to local high-technology development. There is no doubt that this has played a significant role in the area's development, as a special aspect of the "London-focussed" character of Berkshire's high-technology activity. But how far it can be regarded as the dominant influence explaining Berkshire's high-technology growth is debatable. Survey evidence (Hall et al., 1987, 71) reveals that only 10% of Berkshire high-technology companies viewed access to local GREs as a locational advantage, with a 17% value for larger multi-site companies. And very few new firms have originated by entrepreneurial spin-off from GREs, unlike the Cambridge university spin-off phenomenon. Nevertheless, the defence procurement factor does represent another, and important, local environmental characteristic which has shaped the character of this leading high-technology complex, differentiating it from areas such as Cambridge or Silicon Glen.

Silicon Glen

It has been argued earlier that the development of high-technology industry in Scotland primarily reflects the impact of regional policy incentives and strategies on large, multi-plant corporations, usually foreign-owned, at particular periods since 1945. Labour supply advantages in a high-unemployment region have played a secondary role. As table 4.3 shows, Scotland recorded the fourth-best high-technology employment performance in Britain between 1981 and 1984, albeit with a slight decline of 1.9%. And three areas, Fife, Tayside and Borders, were in the highest-growth category of British counties. By 1984, high-technology industry in Scotland employed 86 thousand workers, more than in five other British regions including East Anglia. The main sectors involved are computers, semiconductors, defence electronics and telecommunications, together with some aerospace and pharmaceuticals.

As Sayer and Morgan (1986, 168) point out, central Scotland, marketed since 1975 by the Scottish Development Agency as "Silicon Glen", has often been categorised simply as a low-level branch plant economy, where routine assembly of standardised high-technology products is carried out by low-skilled production operatives in plants vulnerable to closure as labour cost advantages shift to third-world countries. This stereotype is invalid and misleading. While the region is dominated by large foreign-owned companies and plants, and exhibits a relative bias towards production rather than R and D and operative rather than highly-qualified labour (Kelly and Keeble, forthcoming), evolution and upgrading of high-technology activities over the last 15 or so years has significantly enhanced the quality of production and jobs. There has been a considerable increase in R and D activity and local design content, and in new indigenous firm formation. Moreover, as Kelly (1987, 196) points out, the leading US computer companies in Scotland (IBM, Honeywell, NCR, and Burroughs) have all now been operating there for nearly 40 years, indicating "noteworthy long-term stability of these foreign-owned branch plants". The crude "branch plant economy" stereotype is thus not appropriate to Silicon Glen, however much it may fit other UK assisted regions such as south Wales (Sayer and Morgan, 1986).

The evolution of Silicon Glen's high-technology industries has occurred in a series of waves of development, beginning with the war-time establishment of the Manchester-based Ferranti defence electronics company in Edinburgh (Ferranti is now Scotland's leading high-technology employer). Development continued with an influx of US giants in the late 1940s/early 1950s, and with a further surge of American companies in the later 1960s and early 1970s, the latter seeking a European production base in a context of rapidly-growing world markets for electronic products. Since 1975, a new and continuing flow of US and Japanese electronics companies have been settling in Scotland to serve the wider European Community market. Recent major examples include Digital Equipment's £82 million microchip plant near Edinburgh, and Compaq's - now North America's fastest-growing computer firm - £16-million first-ever European production centre at Erskine near Glasgow. This evolution has been accompanied by increasing technological sophistication and R and D content of Silicon Glen operations, with over one-third of the product range of Scottish computer plants being designed there by 1984 (Kelly, 1987, 197).

This evolution, and Silicon Glen's high-technology development generally, have been powerfully shaped and influenced by the nature of the local industrial environment in central Scotland. Early development under regional policy stimuli was confronted by a traditional production-focussed labour market dominated by male-employing heavy industries and skilled craft unions with a history of labour militancy. Not surprisingly, therefore, incoming firms concentrated on assembly operations, which could benefit from capital-focussed regional policy incentives, rather than on R and D, and wherever possible recruited female workers rather than male to greenfield sites such as new towns away from old manufacturing centres.

Since 1970, however, achievement of an essential "critical mass" of high-technology activity in Scotland, the increasing supply of highly-qualified engineers and graduate employees available from Scotland's eight universities and 70 colleges, and a growing need in a period of rapid technological change to integrate research and production at the same location so as to be able to respond rapidly to changing needs in a particular (e.g. European) market, has brought increasing R and D and technological sophistication to foreign-owned Scottish operations. New firm spin-off has also occurred, for the first time, in the high-technology sector since 1975, Scotland recording the fourth largest number of new independent computer firms and jobs 1975-84 in Britain, after London, ROSE and East Anglia (Keeble and Kelly, 1986, 84). Indigenous activity is still however dwarfed by that of multinationals, US companies alone generating 72% of all 1986 revenues earned by Scottish-based electronics plants. The contrast here in nature of high-technology development with local milieus such as Cambridge is thus very striking.

CONCLUSIONS

High-technology industry in the United Kingdom has evolved over the last two decades in a number of different local environments, its scale and nature being powerfully influenced both by macro-level forces and the precise mix of advantages and constraints offered by a particular local area at a particular time. This chapter has demonstrated the surprisingly widespread nature of high-technology growth spatially in Britain, in a context of national high-technology job decline focussed on major cities and conurbations. The latter have of course also suffered from job losses associated with the adoption of new technologies in existing industries, as discussed at the beginning of this chapter. A clear association between expanding high-technology activity and less-urbanised small-town locations has been argued to reflect the key importance to this sector of highly-qualified R and D workers and entrepreneurs, and the marked residential preference of such individuals for visually-attractive areas outside Britain's big congested cities. The existence of historic - or more recent - centres of research activity, of government regional policy incentives, of proximity and communication advantages, and of locally-available production workers, have all played some part in recent trends in different areas. These trends must of course be viewed within a global and national context of rapid technological change, of increasing internationalisation of high-technology industry, and of organisational restructuring including a marked increase in the birth rate of new, locally-engendered, high-technology businesses. But a full and adequate understanding of how and why high-technology industry is evolving in the UK, as in other European countries, does require in-depth analysis at the level of local and regional environments, given the importance of local socio-economic structures in shaping and influencing current development.

REFERENCES

Breheny, M.J. and McQuaid, R. (1987) H.T.U.K.: the development of the United Kingdom's major centre of high technology industry. In Breheny, M.J. and McQuaid, R. (eds), The development of high technology industries: an international survey, Croom Helm, London, pp. 297-354

Butchart, R.L. (1987) A new UK definition of the high technology industries, Economic Trends, 400, February, pp. 82-8

De Jong, M.W. (1987) New economic activities and regional dynamics. Unpub. Ph.D. dissertation, University of Amsterdam

De Jonquieres, G. (1987) Technological change, Financial Times, UK Industrial Prospects, January 5, 1987, p. 4

Freeman, C. (1986) The role of technical change in national economic development. In A. Amin and J.B. Goddard (eds), Technological change, industrial restructuring and regional development, Allen and Unwin, London, pp. 100-14

Hall, P., Breheny, M., McQuaid, R. and Hart, D. (1987) Western sunrise: the genesis and growth of Britain's major high tech corridor, Allen and Unwin, London

Henderson, J. and Scott, A.J. (1987) The growth and internationalisation of the American semiconductor industry: labour processes and the changing spatial organisation of production. In M.J. Breheny and R.W. McQuaid (eds), The development of high technology industries: an international survey, Croom Helm, London, pp.37-79

Keeble, D. (1968) Industrial decentralization and the metropolis: the North-West London case, Transactions of the Institute of British Geographers, 44, pp. 1-54

Keeble, D. (1980) Industrial decline, regional policy and the urban-rural manufacturing shift in the United Kingdom, Environment and Planning A, 12, pp. 945-62

Keeble, D. (1986) The changing spatial structure of economic activity and metropolitan decline in the United Kingdom. In H-J. Ewers, J.B. Goddard and H. Matzerath (eds), The future of the metropolis: Berlin, London, Paris, New York: economic aspects, Walter de Gruyter, Berlin, pp. 171-99

Keeble, D. (1987) Industrial change in the United Kingdom. In W.F. Lever (ed.), Industrial change in the United Kingdom, Longman, Harlow, pp. 1-20

Keeble, D. (1987) High-technology industry and local economic development: the case of the Cambridge phenomenon, Environment and Planning C, Government and Policy, forthcoming

Keeble, D. and Gould, A. (1985) Entrepreneurship and manufacturing firm formation in rural regions: the East Anglian case. In M.J. Healey and B.W. Ilbery (eds), The industrialization of the countryside, GeoBooks, Norwich pp. 197-220

Keeble, D. and Kelly, T. (1986) New firms and high-technology industry in the United Kingdom: the case of computer electronics. In D. Keeble and E. Wever (eds), New firms and regional development in Europe, Croom Helm, London, pp. 75-104

Kelly, T.J.C. (1986) The location and spatial organisation of high technology industry in Great Britain: computer electronics, Unpub. Ph.D. dissertation, University of Cambridge

Kelly, T. (1987) The British computer industry: crisis and development, Croom Helm, London

Kelly, T. and Keeble, D. (forthcoming) Locational change and corporate organisation in high-technology industry: computer electronics in Great Britain. In M. Breheny and P. Hall (eds), The growth and development of high-technology industry: Anglo-American perspectives, Rowman and Littlefield, New Jersey

Lawton-Smith, H. (1987) Technical and information linkages: the case of advanced technology industry in Oxfordshire, Regional Studies, forthcoming

Leadbeater, C. (1987a) Pirelli UK: a painstaking route to job security, Financial Times, February 9, 1987, p. 7

Leadbeater, C. (1987b) KP cracks a hard nut, Financial Times, June 15, 1987, p. 14

MacGregor, B.D., Langridge, R.J., Adley, J. and Chapman, B. (1986) The development of high technology industry in Newbury district, Regional Studies, 20, 5, pp. 433-48

Martin, R. (1987) The new economics and politics of regional restructuring: the British experience. In L. Albrecht and P. Roberts (eds), Regional Policy at the Crossroads, forthcoming

Peck, F. and Townsend, A. (1987) The impact of technological change upon the spatial pattern of UK employment within major corporations, Regional Studies, 21, 3, pp. 225-39

Perry, R., Dean, K. and Brown, B. (1986) Counterurbanisation: international case studies of socio-economic change in rural areas, GeoBooks, Norwich

Preston, P. (1987) Technology waves and the future source of employment and wealth creation in Britain. In M.J. Breheny and R.W. McQuaid (eds), The development of high technology industries: an international survey, Croom Helm, London, pp. 80-112

Sayer, A. and Morgan, K. (1986) The electronics industry and regional development in Britain. In A. Amin and J.B. Goddard (eds), Technological change, industrial restructuring and regional development, Allen and Unwin, London, pp. 157-87

Searjeant, G. (1986) Weak links in twin vision of the future, The Times, Business and Finance Section, November 17, 1986, p. 23

Segal Quince and Partners (1985) The Cambridge phenomenon: the growth of high technology industry in a university town, Segal Quince, Cambridge

Spooner, D.J. (1972) Industrial movement and the rural periphery: the case of Devon and Cornwall, Regional Studies, 6, pp. 197-215

Thompson, C. (1987) Definitions of high technology and government programs in the USA: a case study of variations in industrial policy under a federal system, Environment and Planning C, Government and Policy, 4, pp. 1-17

Townsend, E. (1987) Britain best for US hi-tech, The Times, February 19, 1987, p. 19
Williams, I. (1985) Britain's high-fibre diet, The Sunday Times, October 6, 1985, p. 67

Chapter 5

Local Innovation and Large Firm Strategies in Europe

Claude Pottier

INTRODUCTION

A study of the relationships between small and medium-sized firms and large corporations provides a unique vantage point for considering the nature of recent regional economic development in Europe. If regional development is propelled by small independent firms which exhibit many local linkages and inter-connections, then we speak of endogenous development or development "from below". If in contrast, these firms show a marked dependence on external corporations - which may or may not have local subsidiaries or branch plants that are largely independent of their environment - we characterise regional development as exogenous or "from above". One fairly widespread viewpoint today stresses the logic of an endogenous development that is seen as gradually replacing previous exogenous trends. This assumption is based on a study of the present restructuring of the productive system: a Ford-type production organization that once called for segmented work, carried out in large-scale units, is now giving way to a more flexible type of set-up. The latter is characterised by small units which are more easily adapted to rapidly-changing technologies and markets, these in turn requiring more polyvalent work procedures that more closely and quickly correlate design and manufacturing, thus allowing for innovation. Granted, this focus does underline essential features of modern economic development. But it can lead to two conclusions that are too hasty to be valid and, in fact, have proven faulty. The first of these is the idea that large corporations are being phased out by innovative small firms; the second, that there is a spatial dimension to this supposed trend, in that a more endogenous process of regional development is said to be set in motion by the growth of dynamic, innovative small firms.

We prefer to replace this dualistic logic, with its implication that small/medium-sized firms and large corporations are directly opposed to one another, with a dialectic mode of reasoning that actually brings to the fore the kind of relationships that now exist between smaller firms and large corporations in the innovation process. Dualistic theory often erroneously conflates the small unit and small firm, thus overlooking the strategy of corporate deconcentration, the breakdown of large firms into small units, and their very real control over innovative small firms. In fact, in spatial terms, corporations have shown that they are able to make the most of local resources. The recent mobilization of local institutions in many European regions in order to promote innovative, technological development, and a strengthening of local ties between industry and research, might lead one to suppose that a more endogenous regional development is taking place. In fact, however, we would argue the opposite: namely, that innovations and the technological decisions of local small firms have never been more dependent on external influences - and especially that of large, externally-controlled firms - and that exchanges among local firms in both technologies and goods are in fact generally limited when compared with their outside trading activity.

Thus, our underlying assumption leads us to study local innovations in the light of the relationships between small firms and larger corporations. In this chapter, we shall try to specify the kind of regional development that is linked to the present restructuring of the productive system, bearing in mind that the greatest mobilization of local resources does not inevitably signify a more independent regional development capacity, but the reverse. To support our argument, we shall refer to data obtained by GREMI (Groupe de Recherche Européen sur les Milieux Innovateurs) from a survey of a dozen European regions. We shall also make use of evidence from research in other areas of Europe. Figure 5.1 shows the location of these areas. Thus, we will attempt to delineate the kind of relationships between corporations and small firms in the innovative process in order to explain the exact nature of recent regional development in particular areas.

Figure 5.1 Local Reference Environments: European Case Studies

CURRENT ISSUES IN THE RESTRUCTURING OF THE PRODUCTIVE SYSTEM

Before studying local ties between small firms and corporations, we must first inquire into the implications of the present restructuring of the productive system. In traditional industrial regions, local integration was formerly based on the complementary nature of traditional sectors, characterised by the trading of goods among sectors, the local mobility of a workforce whose skills fitted the needs of these sectors, and technical continuity within the sectors of the same branch. We know that this type of integration has been dramatically reduced, and large corporations with their externally-focussed territorial logic, which has sometimes aggravated the dislocations of local economies, have been blamed for this. Yet it would be misleading to point an accusing finger at corporations in themselves as both the initiators of this trend and the opponents of frustrated small firms seeking to promote local development. Local dislocation has been the result of a global evolution of the productive structure of the capitalist system which is chiefly distinguished by its internationalization, with large corporations as its principal vectors. Today, the reconstitution of the productive system, at both local and national levels, is no longer taking place from a base of sectorial activity but has evolved from the network of ties between research and production, technical and management training (an engineer is now expected to be able to set up his own firm), and innovation and financial circuits (Pottier, 1985). Just where is this new type of integration leading and exactly which roles are the corporations involved playing? These are the issues that must be clearly dealt with.

Non-economic factors in local economic integration

The first point to note is that to a large extent, examples of local integration referred to today in the academic literature appear to be linked to non-economic factors. The most often-cited case of a local "neo-Fordist type" system is that of Emilia Romagna where small units of production are found working together, characterised by flexibility in technologies and employment, polyvalence and job enrichment. Asheim (1984) points out that this model is rarely present in other European countries where traditional society has been severely weakened. Nonetheless, the organization of industrial production based on firm co-operation and job flexibility is evident in other Italian regions, for example in the Marche, though here no new technologies have been initiated. Thus social, cultural, and political factors seem to operate in Italian society to promote local economic integration, a neo-Fordist type system being just one of its models. Stohr (1986) has studied another example of regional development "from below", a co-operative grouping from Mondragon that extends through the south of the Spanish Basque country. It has developed advanced technologies linked to micro-electronics in a variety of sectors. Stohr shows that its success is a direct result of its co-operative structure and control over local political institutions which are fairly autonomous of central authorities. While the grouping works in an open market, it still retains control of its own decisions and its financial network is obliged to honour the principle of reinvesting profits back into the Basque country.

Co-operation among small firms: myths and realities

The important role of non-economic factors in local integration seems conversely to explain why co-operation and interaction between small, innovative firms appear to be quite minimal in the twelve innovative European environments studied by GREMI. The most striking example of this phenomenon can be found in the ZIRST of Meylan, the "Zone for Innovation and Scientific and Technical Projects" located in the Greater Grenoble area where it was taken for granted that this type of co-operation was most likely to develop. Set up in 1974 to enhance local scientific and manufacturing initiatives, this zone today contains 130 firms, most of which are small, innovative micro-electronics and micro-data processing businesses. In their in-depth survey of Grenoble-Meylan, Boisgontier and de Bernardy (1986) emphasize the disparity between the current reality and the initial project proposal which had predicted an increase in linkages and synergetic effects among these small innovative firms, given their spatial proximity. They acknowledge that there are informal networks set up among business leaders or executives from the same university or firm which account for a fair amount of socializing, especially in public places within the zone. But the two researchers also point out that this apparent amiability does not usually spill over into the business world, where firms remain fierce competitors who, when occasion demands, will buy listing shredders to eliminate available but sensitive data or advise their executives to avoid public places. According to Boisgontier and de Bernardy, once this point is reached, the former advantage of "the cafeteria effect" becomes a decided handicap. They conclude that synergy effects in the innovative process can be more readily found in complementary technical know-how than in the spatial proximity of firms within the same industrial zone.

Relationships between small/medium-sized firms and centres for research and technical assistance

A basic feature of what appears to be a restructuring of the local productive system in many European regions is the development of a multiplicity of institutional mechanisms that act as intermediaries between small firms and local centres of research and technical assistance. Yet it is clear that at present, this phenomenon only involves a small number of firms. Thus Peyrache (1986) shows that the impact of St Etienne's potential scientific and technical know-how on small firms is still limited even though the available technical assistance is more or less adapted to the sectorial structure of labour reserves. In this region, "relationships between universities/professional schools and manufacturing are shortlived and involve a small number of local firms", while as for technical centres, "their initiatives basically target big firms, largely ignoring small firm sub-contractors." Many firms, in fact, negatively assess the usefulness of the high-technology production pole that is being set up. In the Swiss Jura, Maillat and Vasserot (1986) affirm that "inter-firm collaboration has still not gotten off the ground and research liaisons are just beginning to take hold there". In Provence, Granie's survey (1983) of innovative firms reveals that they very rarely resort to dealing with innovation-assistance type organizations, either because they are unfamiliar with them or because they feel they are inaccessible, given their bureaucratic slowness and lack of efficiency.

Local resource imperatives for large corporations

Complementarity between large and small firms in the innovation process is a permanent reality in the history of capitalism. Large firms often miss out on crucial innovations because a more or less significant proportion of their activity is devoted to safeguarding their existing monopolistic position in certain markets or technologies, and also because of their need to recoup through depreciation past investment in older technologies. Small firms are not constrained by these imperatives, and can thus often bring about a major innovation in new sectors which require rapidly-evolving scientific and technical skills and where entry financial restrictions are minimal (Keeble and Kelly, 1986; 82: Kelly and Keeble, 1988). The same type of duality can be noted for minor innovations whereby small firms develop specific technologies or adapt technologies to markets they know well. A large firm can itself set up a small, innovative firm; it can also take control of it, come to an agreement with it, or eliminate it.

Due to the instability of techniques and markets, this type of dialectic relationship between large and small firms appears to be undergoing quite a distinctive type of development. Since the late 1970s, large manufacturing corporations have tended to replace their previous strategy of product development and control of clearly-defined markets by one that puts a premium on technology. This reversal has been studied by GEST (Groupe d'Etude des Stratégies Technologiques) (1985) which has proposed the concept of a "technological cluster", defined as a set of technologies that can be used to manufacture diverse products for different markets; each corporation, depending on its size, tries to perfect one or more technological clusters and capitalize on its value in its products and markets. The development and especially the distribution of basic knowledge and expertise calls for a fair amount of centralization. On the other hand, the development of new techniques requires decentralized structures in the form of relatively autonomous corporate units, and the setting-up of a permanent office of technological responsiveness oriented to the corporation's environment, especially public research teams and innovative small/medium-sized firms. Local areas are part of this environment. Each of them harbours traditional know-how and expertise which, once enhanced by the mastery of new technologies, can lead to innovations implemented by small firms in progressively closer collaboration with local research centres. Thus it is crucial for large corporations to be on the look-out for local innovation in order to control it. How tight will this regulation become, and what will new corporate structures be like? Recent case studies provide graphic insights here, in that they all reveal an extra-territorial strategy. We shall attempt to specify just when this strategy can become a permanent phenomenon and then proceed to study how a greater involvement of corporations in local environments is being worked out.

THE EXTRA-TERRITORIAL LOGIC OF CORPORATE STRATEGIES

The extra-territoriality we shall consider is limited by the scope of this chapter, that is, to innovations and technological transfers. It is obvious that corporations almost always have a dramatic impact on local labour markets, which allows them to establish various contacts with local training organizations. In fact, the technological ties of branch plants with their local environments are the weaker, the stronger their degree of corporate integration. Dupuy (1985) makes use of this formulation to define a corporate typology that allows for a study of their relationships in the local environments in the Southwest of France. By order of decreasing intensity of local exchanges, he differentiates non-specialized branch plants, then those that are specialized in one product, and lastly, those specialized in one segment of the production process. We know that this typology largely overlaps with a spatial one: the autonomy of branch plants, the extent of research and innovation, the skill level of their workforce and the number of local scientific and technical exchanges, all decrease in accordance with the size of cities. Well-known examples of branch plants located in outlying European regions that are closely integrated within their corporation yet quite independent of local environments are those in the West of France, the South of Italy (Abruzzi), the North of England, the border between East and West Germany, and the Italian border in Switzerland (Ticino).

This corporate localization process in outlying regions where a largely-unskilled and cheap workforce was available for mass production was typical of the 1960s. In the 1980s, however, new branch plant establishment in such areas appears to be the exception rather than the rule. Numerous researchers have concluded that this trend reversal is not only due to a slow-down in industrial growth, but also to the increasing importance of intellectual work and its steadily-growing interaction in the production cycle. Already in the 1960s, a dual corporate localization strategy based on sectors of activity had begun to emerge. Thus during this decade, both Thomson and Philips located the sites of their general consumer electronics establishments in small French cities but installed their more sophisticated electronics branch plants in university cities. Thomson began to design and manufacture its integrated circuits in Grenoble, which was to become the French pole in micro-electronics, and then settled in Toulouse, a pole of the space industry. In addition, starting in the early 1970s, European electronics groups operating in France began to transfer the assemblage of mass-produced components to the Third World so that the ratio of skilled workers increased in small French city plants. This trend has continued up to the present day in the electronics industry, with increasing automation of assembly operation in European plants. While not systematically committing ourselves to a study of data of corporate location sites based on specific types of plant, such as those characterised by no specialization, single-product specialization, or specialization in one segment of production, it does appear conclusive that the regions which have been studied reveal that the present trend clearly favours setting up branch plants in sites where the entire production process, in the strictest sense of the term, is carried out, with or without an incorporated research programme.

Let us now consider branch plants that are extra-territorial in so far as research and innovation are concerned; that is, those where either these facilities do not exist or where they do exist but have no ties with the local environment. In this context, it is important to differentiate between European corporation activities in their own countries and American and Japanese investment in Europe. Since the 1960s, American investment in the electronics industry has been particularly channelled into relatively large European cities that can provide an adequately-skilled workforce. In fact, the skill level criterion appears to be increasingly crucial in today's world. For example, in 1982 in an effort to take over the European market for electronic connectors, Du Pont de Nemours located its new plant in Besançon because of the high skill level in metalwork there (Pottier and Touati, 1986). Equally, the remarkable development of American and Japanese electronics investment in Scotland since 1969 also seems certain to continue in future years due to the high technical level of the workforce in this region, coupled with its relatively low salaries.

These examples, as well as others we could refer to, all feature branch plants with no fundamental research units but which incorporate the most modern operating processes and equipment requiring a highly skilled workforce. However, we must also include branch plants with a highly developed research capability but few or no ties with local universities. Examples here are the Caterpillar factory in Charleroi (Alaluf, Martinez and Vanheerswinghels, 1986) or the French corporation Merlin-Gérin, which developed its electronics operations in Grenoble without availing itself of research links with the university, which specializes in this very sector (Boisgontier and de Bernady, 1986). These are counter examples of the seemingly growing trend today - the multiplication of ties between large corporations and local innovative activity - whose various ramifications we are now going to study.

SPIN-OFF AND LARGE FIRMS

Spin-offs of smaller units from large corporations can result from two opposite processes. On the one hand, they may reflect a policy of voluntary corporate decentralization; on the other, they can arise from internal conflicts of interest when executives depart to found their own firms. In the former case, an intermediary step in the decentralization process is a decrease in the size of the company's branch plants. In his study of the site locations of corporations in Toulouse, Galliano (1986) notes that since the early 1970s all the plant branches with more than 500 wage earners have suffered losses in overall manpower. This does not mean a slackening of corporate involvement in the local environment, but rather an increase in their total workforce through the setting-up of small and medium-sized units (from 50 to 200 workers). For example, the SNIAS has divided its branch plants into specialized autonomous workshops, while Thomson has divided its subsidiary CET into six units, bringing the number of corporation branch plants in the Greater Toulouse area up to 19. This fragmentation of branch plant activity has brought with it an increase in ties with the local environment.

Let us now look at firm spin-offs that illustrate more obvious local integration. They often materialize in the establishment of subsidiaries to design and manufacture new products. For example, the largest Belgian electrical engineering firm, ACEC, has recently set up at its site in Charleroi three subsidiaries in three high-tech sectors (biotechnologies, office automation and micro-electronics); all of these have major local linkages.

Executive departures that are not corporation-incited are perhaps more interesting examples, for they clearly highlight the complementary relationships which can exist between small and large firms from the initial to the final stages of the innovation process. We have previously explained why major technological innovations may often be developed by small firms rather than corporations. The case of the breakthrough of micro-electronics manufacturing in the United States reveals that a two-fold spin-off process was involved. In the beginning, research engineers left the large firms that had produced the first semi-conductors. Then spin-offs followed from within small firms themselves, particularly those located in the soon-to-be Silicon Valley. To understand this phenomenon, we must not only take into account the relatively endogenous disadvantage of large firms - in that mass production was rarely possible - as the motor-force behind the proliferation of small firms in the growth phase of this sector, but also the factors that led to fierce competition among all firms, small ones included. These factors are the very low entry requirements for production in micro-electronics, and the very promising profit margins (see Planque, 1985, for a study of global reasons). Rather than argue that a local environment, the Santa Clara county, generated the proliferation of small, innovative firms, it is essential to view this phenomenon as an outcome of manufacturing requirements and processes of change which found in this county optimal sites for factory construction, as an extension of the San Francisco conurbation (as its urbanization occurred prior to the development of micro-electronics). As a result, it was endowed with both a concentration of universities and an emerging manufacturing complex (Saxenian, 1985).

The frequently studied example of Silicon Valley is in fact somewhat exceptional because it is the location of contemporary innovations which have revolutionized the world-wide economy. Perhaps a less extreme example is the Greater Grenoble area, the French pole of micro-electronics. It too has been studied in depth because it appears to be the most dramatic example of local expansion by small innovative firms in France. Here again we are dealing with a university environment with relatively strong links between universities and manufacturing firms when compared with the French average. But as in Silicon Valley, the momentum behind the multiplication of small firms can be found in spin-offs from large firms, which in this case are located in the area, particularly the firms of SEMS. Boisgontier and de Bernardy (1986: 193) write: "From a reading of monographs of new firms in the Greater Grenoble area specialized in micro-electronics-data processing, it is quite striking to notice that at least 16 of them have come directly from SEMS; this firm has in fact taken upon itself the role of a real matrix and nurturing mother in the new manufacturing 'micro' network of Grenoble." SEMS was founded in the late 1970s when Télémécanique transferred its computer plant to Thomson. The departure of SEMS engineers (some 50 of them left to start their own firms) was the outcome of a contradiction between highly-qualified research personnel who had developed and designed a new mini-computer and the restructuring logic of a corporation that required them to become competitors of another team and raised doubts as to the usefulness and actual purpose of their work. Following this first series of new firms founded by executives who had left large firms was a second generation of business leaders who were university graduates.

The examples of Silicon Valley and Grenoble allow us to specify the kind of regional development involved, be it endogenous or exogenous. In both cases the initial impetus was the departure of research engineers from large firms, that is, an exogenous regional factor: Télémécanique, for example, was located in the region but had few local ties with it. The endogenous element that led to the development of innovation in these two regions rather than elsewhere was the intensity of local research and its links with manufacturing. The significance of large firm spin-offs, which were succeeded in Silicon Valley by small firm spin-offs, underlines the fact that research cannot promote innovation unless a manufacturing structure already exists. Universities and research institutions in many parts of Europe have in recent years attempted to set up enterprise agencies or innovation centres, or develop other formal procedures to encourage researchers to found their own firms. But it is rare for manufacturing expertise or a manufacturing partner not to have a hand in these moves, especially in the setting-up of engineering companies that design prototypes. Once actual manufacturing begins, that is from pre-mass production onwards, a manufacturing complex seems to be absolutely necessary. Our research on new firm formation in Besançon and that by Decoster and Tabariés (1986) on the scientific pole south of Paris shows that founders of innovative manufacturing firms seldom come straight from public research. These assertions are supported by other work which calls into question the effects of spatial proximity between research and manufacturing innovation. For example, the survey of Overmeer and Prakke (1980) of 50 innovative firms in the Netherlands points to very few ties with nearby research centres.

These arguments indicate the need for a clear delineation of the concept of a technopole, that is a local environment where there is a marked concentration of research activity. We estimate that this concentration is the outcome of very few technical, localization research restrictions which benefit such environmentally attractive and pleasant areas as Santa Clara county or the South of France. The expanding role of research in economic growth does not enable us to conclude that this type of area is in itself innovative. This is only the case if it also possesses an existing manufacturing structure (firms, a certain frame of mind, and financing), as has been the case in the Santa Clara-San Francisco and Grenoble regions. It is therefore wrong to dramatise too greatly the contrasts between this kind of environment and traditional industrial areas which also necessarily broaden their activities to include research, whether this is carried out locally or not (see the typology of innovative environments proposed by Aydalot, 1986). Thus, in all the innovative firms we studied in the traditional industrial region of Besançon, innovation was the end-product of a manufacturing experiment conducted by young graduates with new technologies at their fingertips. The expansion of micro-electronics in the Santa Clara-San Francisco region should indeed be viewed as more of a revolution than an evolution of manufacturing industry. But this is typical of a major innovation. How many examples of regional development are actually related to such major innovations?

TECHNOLOGICAL RESPONSIVENESS

The spatial diffusion of research appears to be limited when it is not linked to an adapted manufacturing structure, though we have argued that in the reverse situation, proximity counts as one of the factors in the restructuring of local productive systems. A careful scrutiny of corporation strategy should allow us further to specify how such systems are articulated. Even when local involvement of corporations is at a minimum, they still maintain links with training programmes and agencies and, in the same way, they focus their strategy on enhancing their overall reputation, even during a period in which they decide to pull out of a region. This may be done, for example, by giving assistance to help found or modernise local small firms. This seeming paradox makes sense if we consider large firm policies in the light of their overall, permanent negotiations with public authorities who play a key role in financing them. In addition, as Perrat (1986) has pointed out, it may be in a corporation's interest not completely to dismantle the manufacturing network when it pulls up stakes. For example, Rhône-Poulenc, after sizeably reducing its manpower in the Rhône-Alp region, not only set up a subsidiary to promote new activities but also established business corporations which, by giving local small companies a hand in exporting, helped to strengthen the network of small and medium-sized firms which it itself needed to maintain and develop it own production ("Rhône-Poulenc Actualités" no 389, 1983). Thus, Rhône-Poulenc wanted to maintain and draw on the continuing support of a local productive structure based on small and medium-sized firms, a structure which in other regions has been dismantled by the large firms on which the region formerly depended. Amin (1986) explores this phenomenon in the Northern region of England where in 1978, three-quarters of all manufacturing employment was concentrated in corporation branch plants, most of which were specialized in the production cycle.

In contrast with earlier trends, we are however now witnessing the expansion of corporate policies that actively seek to safeguard or strengthen the components of local manufacturing networks. It is within this framework that we can best study large firm strategies of technological responsiveness which enable them to make use of local innovations. For example, large corporations are taking advantage of local public laboratories, as in Besançon where the Swiss company Portescap increased the specialization of its branch plant on research in 1971 by employing members of the university's laboratory team in applied mechanics. Another example is given by Sems-Bull in Grenoble, one of whose teams now works out of a university laboratory. In the opposite direction, corporations have also opened the doors of their own laboratories and testing centres to small firms. St-Gobain has a subsidiary whose job is to increase ties with a network of small innovative firms, and three of its subsidiaries are doing just this in the Savoie, Provence and Aquitaine regions. Rhône-Poulenc has followed suit in Rhône-Alps through the intermediary of a university-founded organization. These initiatives have been encouraged by the French Ministry of Research and Technology which has federated them into regional centres of innovation aid to small and medium-sized firms.

Local institutions providing technical assistance can also act to support large corporations. The example of Bosch may illustrate this. In 1979, the firm installed a numerical control R and D centre in St Etienne in order to become better acquainted with the needs of local machine tool firms. Thus, in turn, this corporation expects the local high-technology production pole to allow it to participate in innovative experiments in this sector.

Technological responsiveness also includes financial links and networks. For example, Hewlett-Packard in Grenoble has taken minority shares in local small, innovative firms. Here we touch on a sphere in which many corporations have systematically adopted a policy of working through the intermediary of venture capital companies. The related study that Touati is now carrying out in the Centre-Economie-Espace-Environnement of the University of Paris 1 shows that French manufacturing corporations are involved in most of the major venture capital investments, at national and regional levels. We are referring here to financial arrangements which involve low contributions, but allow large corporations to spot innovative firms, more generally those whose activities might be complementary to their own. Some manufacturing firms have founded their own venture capital companies, such as Elf-Aquitaine which has decided to contribute to the capital of innovative small firms in three sectors, the development of new energy sources, biotechnology, and pharmaceuticals. Olivetti has followed its example in office automation. In sum, regional funding has a definite advantage over national funding as it provides corporations with more detailed and rapid information about local innovation.

THE TAKE-OVER OF INNOVATIVE SMALL AND MEDIUM-SIZED FIRMS

The strategy of technological responsiveness is rooted in the mutual benefits which accrue both to large corporations and small local firms, resulting in an appreciable strengthening of local manufacturing networks. But when innovations are major ones, small firms are often taken over by corporations and their ties with the local environment are weakened or even cease to exist. Thus, major innovations follow a cyclical movement that in the early stages promotes endogenous regional development, but which later on results in a reversion by large corporations to an extra-territorial strategy. Let us consider three interesting examples which support this view.

In the Northern region of England, corporate acquisitions of small firms have been common over the last two decades, affecting a range of medium-sized firms deemed attractive because of the markets they control, together with innovative small firms. For Smith (1986) this strategy has not only tapped innovation resources, but also drained local innovative reserves by eliminating technological transfers among firms. It is not possible to estimate the actual number of local research teams that have been disbanded or transferred to other corporation branch plants outside the region, but it is certain that in most cases local research has become both less significant and less autonomous.

A second case study is provided by Besançon, where the firm of Lip, the local high-technology leader in electronic watches, was destined to become an assembly line unit after its take-over by a Swiss corporation in the late 1960s. Today Sormel, the company that has emerged from Lip, is the most innovative firm in the greater city area, specializing in robotics. Needing adequate financing for its expansion, Sormel decided to forego the funds of a venture-risk company in favour of being integrated into the Matra Corporation which, though it provided needed complementarity, has restricted its research development to short term objectives and reduced the diversity of its activities.

The third example concerns Grenoble. This city was chosen by the Centre pour l'Energie Atomique (CEA-CENG) as the location for its electronics and data processing research laboratory (LETI) in 1963, and this laboratory became the French research pioneer in the field of integrated circuits. Yet, although this centre needed a manufacturing partner to produce these circuits, large manufacturing corporations were not prepared to invest in the area despite earlier promises. So in 1972 the CENG set up its own subsidiary, EFCIS, which collaborated closely with LETI and other local research centres. In overall terms, the CENG has proved to be one of the major stimuli in Grenoble's expansion. Eight firms were subsequently founded by its researchers, and several sub-contracting firms were developed under its auspices. With the LETI, it provided small local firms with heavy equipment which otherwise they would have had no access to. From 1976 on, Thomson progressively took over managing EFCIS, integrated it in its own vetical structure and decreased its trading links with the local area. Thus recent policy-making by two large firms can be clearly contrasted. The former, the CEA-CENG, was not subject to a competitive logic, thus allowing it to develop its fundamental research in integrated circuits. The manufacturing spillover of this type of research was at first dismissed by competing corporations as they saw little immediate profitability in it, a decision that led to France's lagging behind in this key field. Moreover, the CENG was open to its local environment and played a decisive role in its development. The other firm in question, Thomson, was subject to competitive pressures, a fact much more important than whether a company is publicly or privately-owned. Though arriving late on the scene, it was still able to capture for itself EFCIS' innovative capacity by weakening its ties with the local area.

RELATIONSHIPS WITH INNOVATIVE SUB-CONTRACTORS

In addition to their diverse methods of controlling very innovative small firms, corporations have a privileged relationship with their sub-contractors who are engaged in minor innovations or the mastery of new technologies. Volume sub-contracting is not one of our preoccupations here. It is fraught with the difficulty of modernizing its equipment yet, at the same time, is dependent, able to offer corporations very minimal technical know-how. In this chapter we are concerned with specialty and supply sub-contracting. The difference is crucial for the local area as volume sub-contracting often takes advantage of the effects of proximity with a large firm or sometimes only among sub-contracting companies, thus assuring local exchanges. In contrast, specialty and supply sub-contracting is set up as a vertical structure with an outside client. It would seem probable that this exogenous feature of local development is now being strengthened given the expansion of sub-contracting space. Sallez (1981) shows that since the 1970s the proximity constraint has been considerably relaxed as the spatial dimension of sub-contracting is no longer regional or even national. For many western European firms it is now European-wide, with the exception of a few specialized jobs like machine-tooling or surfacing where proximity has proved to be a substantial advantage.

This spatial expansion, which brings with it a much more rigorous selection of sub-contractors (according to Perrat (1986), Renault is currently in the midst of doing away with no less than 1,000 of its 1,600 sub-contractors), points to the latter's technical development and the changing aspect of its relationship with the outside client, as it evolves into more of a partnership. Sallez (1981) points out that the outside client increasingly expects the supplier to share in the design of new products. As for the specialty sub-contractor, his role is to provide technical assistance. With his initial know-how added to specialized equipment, he should invariably keep the client informed of technological developments. Let us study one example of iron and steel sub-contracting. In the Swiss Jura, as in Besançon, a certain number of small firms, confident in their initial competence in the field of mechanical parts for clock-making, reconverted into machine-finishing or fine precision cut-designing for electronics, aeronautics and automobile manufacturing, developing at the same time considerable expertise in articulating new products with the client. Likewise in Besançon, Augé Découpage perfected a manufacturing technique in cut-designing electronic connectors. The SOCOP, another cut-designing firm, is now guaranteed more than half of its turnover from the manufacture of its own products, but it has opted to remain a sub-contractor for this pivotal role in technological transfer allows it to design its own products. It should be noted that these products are usually "general purpose" ones, that is, after necessary adjustments they can be sold in a variety of markets. In other words, the large firm strategy of technological valorization in different markets is grounded in a similar but lower technological level strategy of sub-contractors steeped in local traditions.

Large corporations are thus able to make optimum use of local resources without enhancing local economic linkages. In contrast, it is clear that in Besançon traditional trade relations of cutters and other local firms have been dramatically reduced yet not replaced by contacts with research and technical assistance centres, a distinctive feature of the new productive system which is emerging. Proximity effects are almost negligible for this type of high-technology sub-contracting and the concept of regional space means much less due to the basically vertical structure of technological exchanges.

CONCLUSIONS

The recent evolution of relationships between large manufacturing corporations and small units of different types has involved a splitting-up and greater autonomy of corporate production units, and growing numbers of subsidiaries, new firm spin-offs, and partnerships. These trends might lead one tacitly to assume that regional development is more endogenous today in that small units, in particular small and medium-sized firms, maintain closer ties with the local environment. This theory is supported by the remarkable expansion of local institutions responsible for facilitating synergy effects in the development of new technologies. Yet studies carried out in several innovative environments in Europe have led us to preclude such a hypothesis. It may even be the case that local synergies are so often referred to simply because they have become very weak, at a time when realization of the necessity of replacing old, local synergies, based on intersectorial exchanges, by new ones characterised by links between research and production, has been growing. The development of these new links is a distinctive feature of the current restructuring of the productive system, but we have no right to suppose that this new system is being locally reduplicated. It is more appropriate to conclude that it has found local support. Large firms seek to make the most of local traditions because innovation is a meeting-ground between research, whose purpose is by nature general, and specific production and know-how. They are thus impelled to set up or take control of firms linked to the local environment, and to found technical or financial subsidiaries whose task is to develop technological responsiveness. This strategy involves taking a role in strengthening the local productive system, though this corporate policy is more of a means than an end in itself. When a very innovative firm is taken over, its ties with the local environment are considerably weakened.

We may thus consider the relationships between large corporations and innovative small and medium-sized firms or high-technology sub-contractors as vertical links. They can be inscribed in the dialectical register of events from the emergence to the corporate control of innovation. In the initial stages, the corporation promotes or merely watches the creation of innovative units, as relatively independent organisations embedded in the local environment. It then takes control of these units and terminates or diminishes their ties with the local area. Within these vertical structures, innovative small firms find themselves in fierce competition with one another, especially where they have research contracts with a large corporation. It is therefore not surprising that their will to co-operate is minimized even when they are located in the same zone of activity, as in the case of Grenoble-Meylan. This example also shows that a university environment increasingly fosters the setting-up of firms to the extent that science more directly conditions production, emerging as a "direct productive agent". But innovation can only take root as a product of the encounter of research and a productive organization if the latter is enclosed in a vertical structure largely oblivious to its local environment.

REFERENCES

Alaluf, M., Martinez, E. and Vanheerswinghels, A. (1986) Situation économique, facteurs de redéploiement et innovation technologique; le cas de Charleroi. In P. Aydalot (ed.), Milieux innovateurs en Europe, GREMI, Paris, pp. 163-93

Amin, A. and Thwaites, A. (1986) Technical Change and the Local Economy: the Case of the Northern Region (UK). In P. Aydalot (ed.), Milieux innovateurs en Europe, GREMI, Paris, pp. 129-61

Asheim, B.T. (1984) Informal economy, small firm development and spatial structures in Italy. Paper for the 24th European Congress of the Regional Science Association, Milan, August 1984

Aydalot, P. (1986) Trajectoires technologiques et milieux innovateurs. In P. Aydalot (ed.), Milieux innovateurs en Europe, GREMI, Paris, pp. 345-61

Boisgontier, P. and De Bernardy, M. (1986) Les entreprises de "micro" et la technopole, CEPS, Université des sciences sociales de Grenoble

Decoster, E. and Tabaries, M. (1986) L'innovation dans un pôle scientifique et technologique; le cas de la cité scientifique Ile de France sud. In P. Aydalot (ed.), Milieux innovateurs en Europe, GREMI, Paris, pp. 79-100

Dupuy, C. (1986) Groupes industriels, spécialisation des unités de production et développement régional. Paper for the conference on CREUSET: Milieux industriels localisés et développement économique, Université de St Etienne, May 1986

Galliano, D. (1986) Stratégies spatiales des groupes et systèmes productifs locaux. Unpub. thesis, Université des sciences sociales de Toulouse

GEST (Groupe d'Etude des Stratégies Technologiques) (1985) Grappes technologiques et stratégies industrielles, CPE Etude no 57, May 1985

Granie, P. (1983) L'innovation des les PME provençales, Centre d'Economie Régionale, Aix-en-Provence

Keeble, D. and Kelly, T. (1986) New firms and high-technology industry in the United Kingdom: the case of computer electronics. In D. Keeble and E. Wever (eds), New firms and regional development in Europe, Croom Helm, London, pp. 75-104

Kelly, T. and Keeble, D. (1988) Locational change and corporate organisation in high-technology industry: computer-electronics in Great Britain. In M. Breheny and P. Hall (eds), The growth and location of high-technology industries: Anglo-American perspectives, Rowman and Littlefield, New Jersey

Maillat, D. and Vasserot, J. Y. (1986) Les milieux innovateurs; le cas de l'arc jurassien suisse. In P. Aydalot (ed.), Milieux innovateurs en Europe, GREMI, Paris, pp. 217-46

Overmeer, W. and Prakke, F. (1980) Nieuwe innovative bedrijven in Nederland, Deel 1, Apeldoorn

Perrat, J. (1986) Liens nouveaux groupes internationaux/PMI: quel développement régional? Paper for the conference on CREUSET: Milieux industriels localisés et développement économique, Université de St Etienne, May 1986

Planque, B. (1985) Le développement par les activités à haute technologie et ses répercussions spatiales. L'exemple de la Silicon Valley, Revue d'Economie Urbaine et Régionale, 5

Pottier, C. (1986) L'organisation collective du transfert technologique dans les régions. In J. Federwisch and H.G. Zoller (eds), Technologie nouvelle et ruptures regionales, Economica, Paris

Pottier, C. and Touati, P. Y. (1986) Les conditions de l'innovation dans les régions d'industrialisation ancienne; le cas de Besançon. In P. Aydalot (ed.), Milieux innovateurs en Europe, GREMI, Paris, pp. 247-66

Peyrache, V. (1986) Mutations régionales vers les technologies nouvelles; le cas de la région de St Etienne. In P. Aydalot (ed.), Milieux innovateurs en Europe, GREMI, Paris, pp. 195-215

Sallez, A. (1981) L'avenir des entreprises de sous-traitance en Ile-de-France, CERESSEC, Cergy-Pontoise

Saxenian, A. (1985) The Genesis of Silicon Valley. In P. Hall and A. R. Markusen (eds), Silicon landscapes, Allen and Unwin, London, pp. 20-34

Smith, I. J. (1986) Takeovers, Rationalisation and the Northern Region Economy, Northern Economic Review, 12, pp. 30-9

Stohr, W. (1986) Territorial Innovation Complexes. In P. Aydalot (ed.), Milieux innovateurs en Europe, GREMI, Paris, pp. 29-54

Chapter 6

Technological Clusters and Regional Economic Restructuring

Guy Loinger and Veronique Peyrache

INTRODUCTION

The concept of "technological clusters" applied to the analysis of localised socio-professional dynamics in a context of rapidly changing technological and economic opportunities is a notion that can surely enlighten our understanding of the problems of local economic development. In fact, instead of focussing on the "products of industry" which only represent changing combinations of technical sub-systems, it is surely more interesting to study the industrial technological background, and particularly local expertise and know-how, the latter providing a good basis for optimal adaptability to the market. It is also important to examine the relationship between R and D (Research and Development) and the development of products. The technological approach allows a different evaluation of industrial territorial logistics, compared with non-technical perspectives. In this chapter we will examine the situation from two different angles; firstly, from the point of view of large industrial companies in relation to their territory, and secondly, from the point of view of the complex relationships between local participants in industrial change, whether they be economic, social or institutional. From this perspective, the underlying economic and technological structures of industrial activities are particularly well-defined in certain "territories", due to tradition and to history, and their analysis is a necessary basis for the application of local economic development policies. The main sources of the examples given originate from studies being undertaken at present in the Rhône-Alpes region of France.

The concept of the productive system adopted here is

centred around two elements; research and development (R and D), and the market. Most of the studies carried out in the 1970s concerning the industrial structure and strategies of companies in Europe appear now to be somewhat obsolete. The traditional Fordist model of large companies characterised by mass-production of low or average-priced goods, and by the spatial separation of decision-making centres from Taylorised production sites with their slowly-evolving products, is giving way to other organizational arrangements modifying the structures of the economic landscape. A series of factors have contributed to an important change in the relationship between companies and their technical and productive systems since the 1970s. These factors comprise the ripening and maturing of certain new technologies capable of radically transforming methods of industrial production and of leading to a new generation of products, together with a change in attitude of the economic participants now faced with a rapidly changing labour market. These trends have forced firms and workers to improve their creativity and ingenuity, while at the same time classical structures of social protection are being questioned.

One of the principal aspects of this situation concerns research and development activity, and industrial strategies with regard to what can be called "technological management", as a complement and reinforcement to financial and commercial management. Compared to the one or two-year long strategies whch are normal for production management in a stable context, R and D strategies of large industrial companies usually operate within five to ten-year time-horizons. The central concern of these companies is with pure research and applied research on basic technology; and to facilitate this, they organise different combinations of research activities in order to create a potential store of know-how, a bank of ideas adaptable to economic reality.

This kind of contemporary approach to technology strategy by large multinational companies has led GEST (Groupe d'Etude des Stratégies Technologiques) to propose the concept of a "technology cluster" (GEST, 1985). A technology cluster is a collection of basic or generic technologies, which can be combined or linked together so as to create a coherent and universally-relevant ensemble, capable of application in numerous and different areas of economic activity. Among the most pertinent recent examples of combinations of generic technologies which have led to products with multiple applications may be cited the development of micro-processors, new composite materials, and cloning techniques based on genetic manipulation. Analysis of the behaviour of large industrial firms reveals a recent tendency for many to refocus their activities on mastering and dominating one or several technological combinations, the latter necessitating a major R and D effort linking fundamental and applied research. This is aimed at exploiting their commercial potential to the fullest possible extent, at the global scale, a strategy which simultaneously requires a major capacity for analysing potential demand and very considerable flexibility in the context of current rapidly-changing industrial conditions. In short, it demands great adaptability in the technical-industrial system and in commercial organisation.

The importance of the concept of technology clusters in the 1980s reflects, inter alia, the very rapid rate of current technological change in many sectors, particularly those of computing and data-processing, micro-electronics and new materials technology[1]. This accelerated process of out-dating of products seems however less related to the evolution of "needs" than to the strategies of large companies which use the "technological weapon" to maintain or re-establish a monopoly or to intimidate immediate rivals. In this context, two major trends can be identified operating over the general industrial scene. Firstly, there is a broadening of corporate horizons and research strategies through the development of "technological watch-dogs" and contacts with particular pure and applied research laboratories, as well as through the buying up of smaller, formerly independent, high-technology companies. These strategies afford large companies the possibility of leadership in all the different stages of the "production" of scientific and technical know-how and its subsequent application to industry. Secondly, there is an increasing adaptability to short-term market fluctuations. Large companies are now being faced with new forms of competition, such as newly-invented products, and are responding by showing an increased awareness of market movements. This quest for adaptability has turned the marketing watch-dog into a technological watch-dog, and also brings a new flexibility to the use of production tools, through the development of computer-controlled methods. In fact, the co-ordination of commercial activity with very flexible production systems - through a data-processed conception and management of the general technical production system, which also includes sub-contractors - gives large industrial companies the capacity to respond rapidly to short-term changes in market conditions, with a prospect of very high profits on a range of products best adapted to the economic reality of the day.

This aspect of contemporary industrial change is however only the tip of the iceberg. For most firms, the power of domination is to be found elsewhere, in the mastering of the latest scientific know-how, several years ahead of one's competitors. The concept of technological clusters is a typical expression of this phenomenon. From this fundamentally human potential, especially the intellectual potential of research and development activity, there emerges a real productive force, parallel to the direct physical and manual productive forces, and capable of generating remarkable changes. In particular, we can observe a relative devaluation process of traditional working-class skills and a contrasting revalorisation of intellectual and mental capacities amongst the workforce.[2] The introduction of data-processing into the production process, and the development of computer-assisted conception, do in fact favour a concentration of know-how and productive intelligence during both the embryonic stages and final stages (maintenance) of manufacturing activity. This minimises the role and labour content of production itself: in this stage, human presence is increasingly being reduced to a minimum, generally comprising a role only of surveillance and supervision. But this concentration of know-how requirements at the primary and final stages of the industrial process means a greater dependence on external economic technical methods. This is also true for the largest companies. It is a paradox that maximum long-term potential for profitable production is often the result of maximum initial divorce between research activity and the profit motive. Thus R and D firms located close to universities who are initially engaged in fundamental rather than applied research may well in the long term become the most significant generators of high profit levels and enhanced social status, because their research permits the development of entirely new solutions to old problems.

Because of the great diversity of "real-world" experience, it is difficult to construct a satisfactory model of the rapport of large companies with their environment. In analytical terms, economists have studied successive "generations" of explanations where different factors have been emphasized. Thus debate in the 1960s focussed on financial capital theories, while that in the 1970s concerned theories of exploitation of the workforce. Today, technological theories are at the centre of attention.[3] New divisions are taking place in large companies that reflect the know-how of the firm, its history - and company history counts for a great deal, especially in terms of inertia, the functional compartments that firms maintain, relationships with suppliers, and finally, the nature of the main consumer markets. This process of fragmentation or corporate division is resulting in a new explosion of firm activity beyond previous territorial limits, the main thrust of which is to bring together qualified workers, researchers and managers in locations where these can most readily be found. Each functional element represents a complex or composite sub-system in itself: for example, the car industry is divided into a motor section, a coach-work section, a spray-painting section, and so on. Each section requires great dexterity and specific technology, in such a way that each sub-system considered will in turn possess its own R and D unit, its own methods bureau, and its own production work floor, in the same location. In the same way, sub-contractors increasingly tend, by the use of successive divisions, to reproduce the organisation of the commissioning company.

These trends all result in a very complex inter-industrial network. The effect of the spatial and territorial structure is not easy to analyse in this kind of context, since there is a plethora of possible combinations that co-exist and mingle. One surprising generalization, however, is that urbanized zones and residential areas are once more, perhaps, becoming strategic sites for development. In fact, because of the density of the social and intellectual network of such areas, with all the potential for development which this provides, we can expect an attractive capacity and maximum diffusion of technological and organisational innovations within technologically progressive and rapidly-growing economic territories.

LARGE COMPANIES AND TECHNOLOGICAL TERRITORIES

The above represents a rough outline that will need further development. In particular, it is simplified in that it isolates only one variable, that of technology and the application of scientific research, amongst many others, notably the financial factor, contact with the final market, links with local training institutions and agencies, and the varying residential mobility of the labourforce. There now follows a rough outline of a model of the relationship between technological clusters, large companies and territorial zones.

The first case concerns those large companies which concentrate the whole range of their R and D activity in one area, thus creating the possibility of direct contact with the firm's decision-makers, directors and managers. The proximity in one geographical pole of R and D activity is generally regarded as encouraging and enhancing the exchange of ideas, the stages before and after new product conception (pure research, and applied research and development up to the establishment of a production line), and the lateral bonds between the original basic technologies, thus promoting collaboration within the firm and new product development according to the state of the market. Nevertheless, when industrial production and associated research units are physically adjacent to the company's administrative functions, we can expect different attitudes in the research teams than those we would meet in the vicinity of a university campus, or in a major public research unit, the two most obvious alternatives. In any case, when R and D activities are separated and geographically isolated from production sites, the latter find themselves in a state of maximal dependence upon the intellectual activity of R and D. In this situation, the division of work functions is in fact accentuated by a spatial and territorial division which confirms and reinforces it. Another implication of spatially-separated functional organisations is the hyper-development of telecommunications systems between R and D and production units; advanced communication technology is required, if considerable movement of administrators between production units, research laboratories and administrative offices is to be avoided.

In this context the transmission of data becomes important: the technology of video-screened work is naturally used to the full, with daily video consultations between units on the methods bureau level. At the same time, the sophisticated organisation of long-distance conception systems and especially computer-assisted production increases the number of communication "networks".

One could ask whether it is possible effectively to diffuse technological innovations locally and regionally from this kind of structure: the separation of R and D poles is not favourable to the development of local and regional contacts. However, the sub-contracting of certain functions or specialized activities can favour exchanges. The firm commissioning sub-contract work must of course keep new products and the research on which they are based secret, if it is to maintain its technological leadership. But it also needs to disseminate technology locally to some degree, in order to maintain competition and select reliable partners. Another mechanism for technology transfer operates through the maintenance of equipment. Firms engaged in this type of activity tend to set themselves up within a reasonable radius of a set of similar industrial establishments. This is true of the maintenance of complementary technical operations, such as quality testing and regular checking, which have to be done rapidly. It is clear that in these extreme cases, the effects of territorial technology diffusion is reduced to a minimum. In such cases, the large firm surrounds itself with an aura of local sub-contractors, which practically belong to the company concerned, and beyond the perimeter of which no ripples of technological innovation normally can spread. In this situation a sort of invisible wall isolates the large firm and its sub-contractors from the rest of the local economy. It would not be incorrect to say that these types of companies have erected a "technological barrier" around them.[4] We find them characteristically in advanced technological industries (bio-chemicals, new materials, electronics) as well as in some more traditional areas (food products, the car industry).

The second case or model of the relationship between technological clusters, large companies and territorial zones concerns the decentralisation of know-how by homogeneous technological territorial divisions. Here in contrast to the first case it is more and more frequent to observe multi-national firms whose functional organisation of R and D is closely linked, on the one hand with the firm's productive operations, and on the other with a territorial distribution of each coherent unit necessary to the strategic logic of the whole.[5] In this case of decentralised R and D, the large companies involved are motivated by several factors. These include a certain degree of complexity, and certain pernicious attitudes resulting from a juxtaposition in one region of all the R and D units. The latter may lead to bickering and rivalry between research teams, involving guerilla warfare in order to obtain preferential treatment from management when new projects have to be chosen. A general burden of administrative procedures may inhibit research initiatives and encourage a state of inertia, with projects being abandoned under the sheer weight of bureaucracy. With a decentralized system, then, the firm's research strategy puts the emphasis not on combining traditional know-how already available within the firm, but rather on newly-emerging combinations that reflect joint consultation between researchers of different origins, with the outside world, and between researchers and production workers.

This second model thus involves both a team of engineers who design the prototypes, and a local group of fundamental researchers regarded as leaders in their own field by the firm's management, working together in relation to a specific commercial activity with its own territorial scale of operations. This combination can create homogeneous, localised, technico-productive divisions. In fact, the localisation effect often correlates with a desire expressed by the administrators to impose their own strategic choices. To choose a new territory is part of industrial strategy and is a stage in the development of a company. It means a risk has been taken, the industrial adventure accepted: much thought usually goes into such decisions.

One of the interesting aspects of this type of functional and spatial strategy is that it involves in certain cases a tightly-articulated bond between research laboratories, units developing prototypes, and those engaged in industrial applications and mass production. This has the advantage of allowing rapid feedback between the different sectors, since the different teams are virtually on the same site. Furthermore, when the life cycle of the product is short, as is often the case in micro-electronics, changes to the technological system can be made quickly. In some cases, all stages of research, production and marketing, even including distribution to the world market, are carried on together, though always as part of the overall company strategy. Such an approach does however work best when confined to one category of products, or even single products, at a given location.

In this model, then, the dominant effect is not a horizontal division (white collar workers and blue collar workers) but a vertical division based on a functional logic. Technological territorial segmentation results from a hardening of the core, through strategic management of the whole. In this system, only a small number of administrative staff are needed to control the whole company, as each isolated segment has no chance of surviving on its own. Technological co-ordination in this case occurs at the top for all the main strategic activities, each unit having just enough scope and autonomy to realize its potential at the local level. We can imagine internal competition by situating on a sphere two separate points of the same nature, each point exercising monopolistic control over its immediate surrounding surface, but competing only on the fringes.

Obviously, with this type of company structure, opportunities for local external technological and scientific contacts will be very well defined. In this case, then, the "game" will consist of mixing with the local environment as much as possible. In this context, any managing director of an average-sized firm must consider at least two important factors. One of these is scientific research and the use of local "brain-banks" such as universities and public research laboratories; the other is the employment of local sub-contractors, even if this means releasing considerable information. With sub-contracting, however, information is usually functionally divided, so that the sub-contractor remains very dependent on the commissioner and will have little chance of escaping from this situation. Bearing this in mind, there will be a tendency for a bank of ideas to grow in the neighbourhood of the commissioning firm, and new qualifications will develop in contact with this milieu. Potentially, this could generate a separate R and D unit; but it generally benefits and is focussed upon the large company, which is able to accept and reject ideas and people as it wishes and according to its needs.

Selection of the site of a decentralised unit, which is to integrate all stages of production from R and D to final manufacture, is a crucial decision. If the choice is good a flow of investments should be of benefit to workers already qualified in any given function. The notion of a pre-existing pool of qualified workers is important because it demonstrates the value of the stock of local industry that has created the local labour environment, the network of expertise in the particular industry involved. Central and local government authorities are in the same position, in their "primary accumulation" of R and D in a particular industrial sector, and in their assistance to large companies in choosing a site to install a new complex unit of production. This of course also carries with it the risk of subsequent economic problems should the firm fall into difficulty. Nonetheless, both local government and workers are likely to benefit from such an installation as it helps to highlight the technological competitiveness of the area chosen; they share the same professional ideology. It is however important to keep in mind the limitations of this fragile kind of economic development, because such decentralised units are of course dependent on wider corporate R and D strategies external to the activity area. Whereas the first type is more characteristic of traditional multinational companies, the second type is more common amongst more recently emerging industrial groups. But we should not oversimplify a complex reality. Whilst the Apple company concentrated all its research units in Santa Clara in California and then set up computerized production facilities in South-East Asia, IBM has research units all over the world, including laboratories at Montpellier and Corbeil in France, and all these branches co-operate with the company's overall technological plan. A prototype of the hyperconcentrated variety is provided by the case of Philips, which has its head office, board of directors and R and D clustered together at Eindhoven in the Netherlands. Lastly, in Grenoble, Hewlett-Packard is an example of another prototype firm which operates a complete production unit manufacturing calculating machines.

LOCAL STRUCTURES AND TECHNOLOGICAL CLUSTERS

The historian Bertrand Gilles (1977) has clearly shown that in order to grasp technology as concept, the determinants of any particular technology have to be understood along with its historical, social, economic and even philosophical origins. Whereas scientific progress is often unrelated to its historical context, the history of techniques as described by the ancient Greeks reveals the opposite; that is to say, that "materialisation of techniques", to quote the expression of Simondon, "cannot be isolated from its soil of origin" and that its social history is connected to this "materialisation". In this context, it is noteworthy that recent years have witnessed a revived interest in what we call "Industrial Archaeology", which holds a great appeal to the general public. This approach to techniques in terms of "milieu" is central to the analysis of the relationship between local and regional territories and the problem of technological innovation. In fact, we too often forget that before being connected and inter-connected by new modes of transport resulting from the industrial revolution of the 18th and 19th centuries, most European regions lived essentially closed, internally-focussed, lives, with the exception of coastal zones and regions irrigated by large rivers.

What is true for agriculture is perhaps less so for industry. Nonetheless, examples such as the development of the metal industry of the Creusot region close to the mines of Montceau-les-Mines can be explained largely by the impossibility of transporting any distance the coal for the manufacture of coke necessary in the production of cast iron. Often, however, it is the presence of an abundant workforce very strongly attached to the land and to its local area which explains the specific local know-how; thus the historical development of local craft skills in many European agricultural regions reflects the time which was available to peasants for such activities during long winter evenings. The present-day plastics industry of the Oyonnax region, for example, grew out of an earlier craft tradition of manufacturing pipes, combs, and box-wood carving. Another case is the establishment of the electronics industry in Grenoble, which reflects a variety of factors, including the creation of a hydro-electric plant to produce hydraulic energy in the beginning of the 19th century, and the existence of an alpine social culture which stresses independence and self-reliance, personal ingenuity, and a great pride in well-finished work. We can recall the existence of a once-flourishing lace industry in the valley of the Arve in Savoie, which now harbours an electronics industry operating in rural work-shops. Similarly, in the textile industry, it is no coincidence to find the silk-spinners of Lyon prospering today as part of a high-class textile industry, which includes firms such as Brochier, now owned by the Swiss multinational Ciba-Geigy.[6]

Many more examples could be quoted that demonstrate how structured know-how in given areas and at particular periods has evolved and enabled economic restructuring, as a response to changes in the wider economic, technological and social environment. It is interesting to study the modes of adaptation of differing localised technical cultures. In certain cases, the latter are the result of the slow, long-term evolution and stratification of local social practices, marked by power struggles over methods of production, and characterised by ethical, individual, attitudes. In others, they are the product of confrontation with an external, foreign, technological and production system, which is thus considered dangerous, destructive, and disturbing but at the same time a possible opportunity for a new start, a possibly enriching experience, that will bring a new freedom. The latter involves a complex process that only occurs in certain conditions, notably during periods of crisis which encourage an open-mindedness towards ideas of substitution. But often the crisis - for example over declining sales - has to be very obvious or even acute before foreign techniques, such as the buying of equipment, foreign travel on technical and professional missions, or the creation of schools for commercial training, will be adopted. Thus the local technological cluster is created, made up of successive strata, rejections, mistakes, and integration of elements into the main structure. Is it possible from these observations to develop one or more general models of a local or regional technological cluster?

The first case concerns a technological cluster built around small and medium-sized firms within a particular industry, often with a long history of local specialization. Examples of this kind are numerous in Europe: they include towns like Prato in Italy, noted for the promotion of ready-made articles, Stoke-on-Trent in England with its reputation for pottery and ceramics, Mulhouse with cotton fibres, Roubaix with cotton and wool textiles, and so on. In this kind of cluster, urban structure helps to reinforce economic specialisation and development. The town acts like a sound-box which amplifies and stimulates the local economy. The local society feels encouraged, protected in this industrial venture of their own making based on their own know-how and abilities. The community acts here as a safety net for individual members of the group. It encourages a communal solidarity, and the exchange of ideas, opinions and information.[7] It is also interesting to note the related activities which often graft themselves onto the primary industries. Thus at Oyonnax, the SMTT Billion Company manufactures plastic moulds for the local plastic injection industry, while similarly-linked activities are associated with the textile industry in Mulhouse, and the pottery industry of Limoges. A final but important aspect is the image projected by such a territorial cluster, the collective image perceived by the outside world, the region that plays the role of an advertisement: "I am here, in this area, this is my field of action, I am indispensable..."

In such regions, periods of economic crises and recession can be sometimes overcome by internally-generated staying power. But this type of milieu can also become an obstacle to change and essential evolution of traditional crafts and techniques. Local know-how is only an expression of the social relationship to production: in order to leave roles unchanged, a certain status quo may persist in methods of production which can bring about a weakening of local managerial and professional capacities, held back by the traditional habits and customs of the region. The development of a new substitute product or different working methods linked to a different social context can result in the collapse of a traditional regional industrial economy of this type within only a few years, and its transformation into an anaemic zone incapable of reacting to external challenges, because of its dependence on traditional methods. In this context of weakening of the social fibre, it is not unusual to witness the arrival of a foreign investor, offering some degree of new employment, based however on this company's own external markets, its own suppliers, and its own equipment, production and personnel management. This is dramatically epitomized by the recent example of buddhist monks entering a Vosges valley to bless a "lost" factory regenerated thanks to Japanese investment. The good genie, and generous provider of a few local economic crumbs, can sometimes, and perhaps all too easily, vanish like a ghost a few years later if it finds a better location elsewhere. But it is also true that regional economies can regain new vigour from a healthy branch unit which has been successfully implanted within the local milieu. Here the concept of technological clusters is particularly useful, as an approach to describing the passage from one state to another.

In contrast to small-firm economies, the second type or model of technological cluster is that built around one or more large industrial companies. Examples of this type are numerous because they are part of the 19th-century industrial heritage of many European regions. One category is the location of highly-structured industrial units close to headquarter administrative offices. Examples here are provided by the West German and Italian car industries, whose major firms and their integrated activities are of key importance for particular local and regional economies. But when, as is often the case in France, management, brain power and key functions are hived off and located in major core regions such as Greater Paris, remaining provincial establishments are more or less beheaded, and lose decision-making powers. From then on, if large companies destabilise their peripheral branch plants or subsidiaries, it is difficult to control the situation in terms of local employment decline, as we have seen in many older industrial areas (Loinger and Boissel, 1986). When large local firms find themselves in difficulty and hence no longer able to provide work to sub-contractors, the effects on the local environment may be cataclysmic, as for example in Dunkerque and La Seyne around the ship-building industry. Such crises may generate manifestations of solidarity in the form of local or regional government intervention, including designation for regional policy assistance, as in St Etienne. This area has witnessed the creation of a new economic and social network stemming from a coming together of qualified experts in the region. Such regions can become an experimental laboratory, a fertile seedbed so to speak for new initiatives, which can sometimes improve the situation. The case of the Creusot-Montceau-les-Mines zone is a good example here. This is the focus of a recent initiative by the region of Burgundy (Bourgogne) entitled "Bourgogne Technology". The latter involves a new Regional Office for Technological Development (ARDT) which brings together Dijon University, Framatome, The Atomic Energy Commission, and the Technical University of the Creusot region in order to create a bipolar centre for Dijon-Le Creusot called the Regional Centre for Innovation and Technology Transfer. This is developing regional expertise in high-performance welding, using high-quality laser technology. This initiative is thus trying to combine traditional know-how with 21st-century technologies.

A third case or model concerns local technological clusters created by the combination of two or more technological systems. Historical evolution may reach a point at which various crafts and know-how of totally different natures take on new significance and vitality. Thus the area of Lyon, for long considered a sleeping beauty, suddenly finds a new energy and vitality because of new combinations of technological, scientific and human factors. The example of Brochier, a firm which operates at the cross-roads of the region's chemical and textile traditions, is very representative of the general situation. Local large companies have created structures which support and stimulate small and medium-sized firms. A forum has evolved which allows Lyon's citizens to exchange ideas in a sort of Brownian local motion enhanced by the closeness of the city's inhabitants, and the informal manner of the gathering. This new system allows small firms to benefit from the experience and technical equipment of the region's large industrial companies (CREATI). A run-down area of Lyon will soon be the site of the "Ecole Normale Superieure", new R and D firms, and a new residential rehabilitation programme. A successful economic future for Lyon can indeed be predicted with some degree of confidence, based on the three key local industries of materials, chemicals, and data-processing. Lyon is also a leading centre of education, research and training, with its University, the INSA, Central School, the Higher School of Commerce, and other institutions. In fact, Lyon is still a city of human proportions (one million inhabitants), while the two-hour high speed train journey to Paris (T.G.V.), the nearby airport of Lyon-Bron, and the proximity of Geneva are all factors in favour of development.

More generally, in the French case at least, changing social attitudes and ideologies focussed on aspects of individual freedom, combined with a Renaissance image of a proliferation of small, high-technology firms capable of earning enormous profits as in the USA of the 1970s, has generated a favourable climate for the creation of small firms originating from university and research environments. The latter often possess close contacts with large public research laboratories, yet are also independent of them. Such possibilities have been co-ordinated by far-sighted municipal representatives such as Mr Dubedout of Grenoble, as well as by local government officials and industrialists like Merlin Gerin. Their initiatives have helped particular towns to flourish and develop new activities, as for example with the ZIRST (Industrial Zone of Scientific and Technological Research) at Meylan near Grenoble. Only a few years after its creation, the ZIRST, which now contains over 100 market research bureaus, micro-electronic prototype manufacturers and inventors, robotics firms, and other high-technology enterprises, is proving to be a great success. The ZIRST seems to be an extremely adaptable tool, allowing for all sorts of social "games" through a flexible organisational process. The list of occupants varies constantly, living proof of the mobility of its structures and population. In short, the ZIRST seems to provide an environment which stimulates the development and creation of a wide range of activities, ranging from printed circuits to specialised manufacture of equipment for large company clients on the local and regional levels, as well as on the international scale.

In the ZIRST example, three scales of contacts may be identified, which result in a network that is dense, yet mobile and constantly-changing. These three scales are: 1) Networks between designers for the development of new products that require specific co-ordination of know-how. Some of these will come into existence in the ZIRST, others will be produced by companies operating both inside and outside the ZIRST. 2) Networks between ZIRST companies and commissioning agents from large research centres like the CNET and the CENG situated in the proximity of the ZIRST. These centres also operate laboratories elsewhere, as for example with CNET at Lannion and Issy les Moulineaux. 3) Networks between small firms within the ZIRST and linked high-technology sub-contractors usually situated within a regional radius corresponding to a "same-day" return car journey of 100 to 200 kilometres. The proximity of these linked firms is therefore not that of a neighbour, but resembles rather the by-gone valley organisation of Alpine Europe where business was carried on from valley to valley by different mountain communities.[8]

In these ways, in areas such as Lyon and Grenoble, complex networks intertwine giving a characteristic profile of a region through a process of growth of innovations around scientific and technological poles. Though geographically concentrated, these allow a very large number of combinations of economic, social and technological factors in a given field of knowledge.[9] This does however require a strong and continuing regional commitment to technological and economic progress. Thus a potential crisis which exists in Grenoble because of the vulnerability of its electronics industry is being held at bay by greater diversification of its R and D activities.

CONCLUSION

The link between the two parts of this study remains to be discussed; that is to say, how may the large industrial corporation's view of localities or regions be reconciled with that of an area's own social and economic participants, institutions and firms? There is much that could be said about this relationship; but one possible operational approach, which could be developed as a professional activity, involves what may be called "territorial technological management". It has in fact become increasingly important in the Europe of the 1980s for government or other institutional intervention to create clustered economic and technological zones, whether they be local or regional. This approach can both help protect the local industrial heritage and maintain historical know-how and expertise, and enable an optimal realisation of regional economic potential in the present-day context of an increasingly global economy.

NOTES

1. On the other hand, in biotechnology, the research and development process for new products takes between five and ten years before actual manufacture can begin, because of the degree of specialization and technological complexity of this new sector.
2. What may be called "direct intelligence", gained over many years of apprenticeship and through many generations of working-class culture, still remains a valorised social standard, but its professional basis has been weakened.
3. In spite of the fact that our "great ancestors", and especially the classical economists including Marx, always stressed the technological element of their theories.
4. This is particularly true of aeronautical sub-contracting companies in the Midi-Pyrenées region focussed on Toulouse. It must be stressed that there are of course two types of sub-contractors, the elite, and the rest (INSEE, 1985).
5. The following remarks are based on research on Hewlett-Packard currently being conducted by the present authors.
6. For further information on this point, consult the publications of Jean Saglio of the university of Lyon (e.g. Les systemes industriels localisés, GLYSI, University of Lyon II, Rapport pour le programme Science, Technologie et Société du CNRS.
7. In the assisted regions of France, the government agency DATAR is attempting to resolve this problem by external intervention; but in most cases, success is limited.
8. For further information on the development of

technopoles in France, with particular reference to the industrial transformation of the Grenoble region resulting from technological innovations in micro-electronics, see Boisgontier and De Bernardy (1986).

9. These ideas are currently being elaborated further in the research project entitled "The integration of new technologies into industrial structures" being conducted by the present authors for CESTA/Commissariat de Plan.

REFERENCES

Boisgontier, P. and De Bernardy, M. (1986) Les entreprises de micro-electronique et la transformation du territoire industriel grenobloise sous l'effect de l'innovation en micro-electronique, CEPS, Université des sciences sociales de Grenoble

Camagni, R. (1988) Functional integration and locational shifts in new technology industry. In P. Aydalot and D. Keeble (eds), High-Technology Industry and Innovative Environments: the European Experience, Croom Helm, London

GEST (Group d'Etude des Stratégies Technologiques) (1985) Grappes technologiques et stratégies industrielles, CPE Etude no 57, May 1985

GEST (1986) Technological clusters and the new industrial strategies, McGraw-Hill, Paris

Gilles, B. (1977) History of techniques, La Pleaide, Paris

INSEE (1985) Cahiers de l'INSEE de la Région Midi-Pyrénées

Loinger, G. (1985) La diffusion des innovations technologiques, La Documentation Française

Loinger, G. and Boissel, J-F. (1986) The technological development of small and medium-sized firms in Burgundy. Research report for the Regional Committee of Burgundy

Chapter 7

New Technologies, Local Synergies and Regional Policies in Europe

Jean-Claude Perrin

INTRODUCTION

In the first stage of its life-cycle a concept emerges as an intuition of a fruitful direction of research. Such is the case with authors like Andersson (1985) and Stohr (1986a) when they refer to "local synergies" for explaining regional creativity in the field of new technologies.

As Stohr (1986a: 16) points out, quoting the Encyclopaedia Britannica, "the concept of synergism is taken from chemistry and pharmaceutics where it denotes that the effect obtained from the combined action of two distinct chemical substances is greater than that obtained from their independent action added together". Thus, when analysing technological innovation, the concept of synergy helps to focus on an essential topic: why and how are relations between participating factors creative and how may that creativity be increased? Behind the notion of local synergy stands the hypothesis that decentralized territorial structures - not only proximity advantages but the whole regional organisation - may constructively contribute to industrial innovation processes. But how is that so?

In his research monograph on "Territorial Innovation Complexes" Stohr (1986a: 17) develops an interesting argument in terms of mutual interactions: "... not only the presence of specific agents/institutions within a region but their mutual interactions is a prerequisite for optimizing regional creativity and innovation..." Nevertheless, that type of reasoning leads to an explanation of local innovative capacities based on transposing to a sub-national (territorial) level, the synergetic properties of "functional-sectorial" processes (the effects of mutual interactions between agents characterized by their functions - research, production, marketing, etc - and by their sectors of activity) which macro analysis is dealing with on a national level. In contrast, it may be argued that the real interest of the "local synergy" concept is that it emphasizes the properties that territorial structures exhibit concerning the improvement of industrial processes.

More generally, one must appreciate that innovation is both a functional-sectorial and a territorial phenomenon and that research has to be developed at the interface of industrial and of territorial processes. But since the territorial aspect has, in the past, been ignored by micro and macro approaches and biased by traditional Regional Science methodologies (Perrin, 1987), a special effort has to be made in order to (re)introduce it. Our attempt to develop the concept of local synergy stands in that perspective.

One may compare local synergy which relates to development questions, with the notion of externality that Alfred Marshall created in order to characterize the impact of territorial environments on the current functioning of firms. Both terms deal with effects of territorial structures on industrial processes. Unfortunately Marshall's intuition appears to have been strongly biased by subsequent neo-classical analyses (Catin, 1985). Micro and macro models cannot capture the spatial dimension which lies at the root of "territoriality". Thus, we prefer to use a meso approach. Technological innovation will be viewed, in this chapter, as a collective phenomenon which must be analysed as an organization and in terms of a network: the nodes comprise the micro actors (institutions, agents...), with the global network forming the macro structure whch is segmented through systemic procedures.

In the first part of the chapter, we analyse the territorial component of innovation and emphasize the strategic role that it should play in the present phase of technological ruptures. In the second part, we try to show how the analysis of local synergies might be applied for improving regional technology policies.

THE TERRITORIAL COMPONENT OF TECHNOLOGICAL INNOVATIONS

In order to be more concrete, we will first consider the new collective innovation processes which have recently emerged in the field of high-technology industry and demonstrate their territorial nature. Then we will deploy the analysis on a more general level and in a more systematic fashion.

The territorial content of new efficient modes of high-technology innovation: from incubators to technopolises

Various authors (De Bernardy and Boisgontier, 1986; Rispoli and Volpato, 1986) have shown that due to the information revolution and micro-computer technology, small and medium-sized firms (SMEs) have developed new innovative capacities. A threshold has been passed so that SMEs are now becoming significant originators of major innovations.

On the other hand, along with the growth and maturing of "science based industries" (Pavitt, 1984), research laboratories which associate fundamental and applied work are playing a leading role through what is commonly called "incubation". Whatever the type of spin-off (either from laboratories or from large enterprises), small firms are created and develop around those academic institutions. Even if, because of their flexibility, small structures are better suited than large ones to face the unstable conditions (technology, markets...) which prevail in high-technology sectors, they are nonetheless often structurally dependent for their development on technological and organizational services. Spatial proximity appears as an efficient solution to that functional constraint because it is an informal and non-restrictive one. That is the reason why innovative small firms prefer to locate in the neighbourhood of incubators and in places where the different highly specialized services that they require are easily accessible. The clustering of these complementary activities in the same area is a matter of territorial organization.

The experience of Sophia Antipolis shows how appropriate territorial structuring helps to extend technological performances far beyond incubation processes. In mixing incubators with R&D and multinational high-technology production establishments, innovation capacities may be greatly enlarged (Perrin, 1987).

Thus, the new territorial innovative structures which have emerged and developed recently are a significant example of local synergies in the field of technological creation.

We have, now, to develop a general explanation of the phenomenon. It is to be found in the specific properties of territorial structures.

The general properties of territorial organisations regarding innovation synergism

Viewed as collective processes and in the perspective of synergism (i.e. of creative interrelatedness), innovations may be conceived as proceeding from **connections between complementary activities** (research, production, marketing...) and sectors (or technologies) or, more exactly, between the corresponding **actors** (institutions, agents) and the **networks** which are activated for their operating. In a "pre-innovative" situation, these components are acting separately as a consequence of their specialisations, and because of competition, of traditional antagonisms between private and public sectors, and of the specific inbred cultures that institutions and networks generate. These constitute structural constraints inhibiting the development of connections. Moreover, innovations are destructive (-creative) processes. They generate ruptures within institutions and within the prevailing development networks which therefore react all the more strongly. That is why innovative connection development is always problematic. It can only be promoted through an appropriate organizational structuring.

We shall first characterize the innovation organizational system which prevailed during the "glorious" decades of growth in the 1960s and early 1970s. We shall, then, emphasize the fact that, in the present phase of technological ruptures, other forms of communication settings are required and, finally, we will show that territorial structures exhibit organizational properties which fit those requirements.

Innovation systems and multinational firms

During the growth period, multinational corporations were the leaders of generic technologies. The process of technological innovation was often powerfully controlled by means of formal institutional structuring. New specialized development functions (R&D, educational training, long-term financing, international marketing...) were created and set up within the corporations themselves in the form of new departments, which, later on, were often transformed into separate establishments. The multiple inter-relations between these departments and with more routine activities were formally structured. In most cases, centralised decision procedures were developed in order to bring more coherence into that complex system. Thus, synergies proceeded through internal, formally institutionalized channels and networks.

As far as spatial arrangements are concerned, multinationals were practising the "spatial disjunction" (Planque, 1985) of their establishments: they (de)localised them within the most advantageous resource environments, depending on the functions and the sectors. Local synergies were only acting on that level. Creative communications mainly proceeded through the closed trans-territorial channels internally organized by the corporate structures.

During this period, advanced research was basically conducted within academic institutions (whether private or public) which were rather loosely linked to industrial units.

As far as small and medium-sized firms (SMEs) are concerned, sub-contractors were technologically directly dependent upon large companies, while autonomous enterprises contributed to the development of technologies (mainly through product development) but not to their creation.

Finally, in that system, conglomerates were in control of major innovations. They decided, in a more or less oligopolistic fashion, about the opportunity of exploiting their own inventions as well as public scientific discoveries.

New innovation imperatives

Changes are, nowadays, occurring on the innovation stage. New actors are appearing: small and medium-sized firms, academic institutions, incubators. Multinationals themselves are looking for new strategies.

The high risk of investment in high-technology R&D activities, the intersectorial nature of new project studies, the shortened life-cycle of products, have had the result that even large firms are no longer able to generate, on their own, the technical bases necessary for their growth (Mariotti and Ricotta, 1986). They are therefore forced to move from the traditional "in-house generation" of research and development to co-operative procedures with other firms (large and small), and academic institutions (public and private). Such strategies are accompanied by organizational changes. Conglomerates decentralize. They set up small autonomous units which are encouraged to promote direct agreements with government laboratories, private and public R&D consortia, and small research organizations, as well as with other companies.

Territorial structures as new innovation environments

With more institutional actors of different types, and more open and flexible strategies, innovation, in the present period of technological rupture, requires new forms of network structuring. We shall demonstrate that local systems display specific properties which fit those requirements, since they are, by nature, places of highly intensive inter-functional, sectorial and social communication.

In the first place, we must emphasize that local synergy is not simply a matter of proximity, as experience proves. Let us take, for instance, the southern zone of the Parisian agglomeration. This contains probably the largest concentration of research and high-technology production activities in France. Yet cross fertilization and local synergies within the area have up to now been rather marginal (Decoster and Tabariés, 1986).

We must therefore look deeper into the very nature of territorial systems. Their finality is that settled populations live and develop harmoniously. In order to reach that goal, a very high level of organization is necessary. Let us recall, for instance, that in cities and regions, "local" activities which serve the local population (in advanced countries, these are extremely specialized and diversified) are tightly interconnected through complex networks whch have been disentangled by "circuit" analyses. These local circuits are articulated to "export base sectors", so that the whole set is strongly integrated (Perrin, 1974). Moreover, this form of economic structure is associated with a spatial organization (spatial division of functions and sectors, urban hierarchy, transport network...) in such a way that local systems are places of multiple and open communications. The informal meeting and melting process is particularly intensive in the centres of cities. These provide a good example of the general intercommunication property of territorial systems.

For all these reasons, local structures constitute a good "substratum" for the setting up of multi-partner communications and for erasing inter-functional, institutional and cultural barriers.

They also play a role comparable to that of an **"informal institution"**. The latter may be defined in the following terms. In socio-economic systems, institutions represent an instrumental framework by means of which greater continuity, stability and coherence are brought into human relations. These attributes are important for innovative collective projects. But normal institutional structures suffer from a serious weakness: their formalism is a factor of rigidity and we have seen that, in the present period of technological rupture, the new forms of innovation require open and flexible modes of communication. In a territorial context, the fact that the many partners are located and settled in the same area is an element of stability and of durability for their common relations. More than that, the communication network is quite open: new information and know-how can enter and complementary co-operation is available whenever it may be needed for technology clusters. More informal networks confer greater autonomy on their members, and autonomy is the most essential factor of creativity.

Thus, the **territorial substratum** brings with it the capacity to associate connexity with independence; that is, to achieve the subtle balance on which high-technology innovative networks are founded.

Since creativity proceeds from interactive impacts of complementary know-how, institutions should integrate external agents for fairly long periods. The mobility of personnel is a condition of efficiency for innovative structures. Aside from institutional aspects, the decision to move into another structure is easier to implement within limited areas.

A further consideration in terms of creativity constraints is the fact that, for operating risky projects, **trust** is as an essential factor. In the field of joint ventures, the latter has been often assimilated to a "capital good" (De Bernardy and Boisgontier, 1986). Territorial organizations may contribute to its accumulation in two ways. During the preparatory phase, reliable topic information concerning possible regional partners can be rapidly mobilized by interrogating the various professional and social networks and cross checking their opinions. At a regional level, social control may be properly and conveniently operated. Later on, in the development phase of the venture, the sharing of the same environment (cadre de vie) helps strengthen personal relationships and, in so doing, reinforces links and commitment between regional partners.

We have successively analysed the connecting properties of territorial structures and their specific contribution to the creativity of communications. We turn now to more dynamic aspects. Innovative territorialized systems such as incubators and technopolises have polarizing effects. Not only do they attract, from outside, complementary activities but the establishments which grow up within them always localize their operations in the neighbourhood, so that the creative network is reinforced and the whole innovative capacity increased. Through territorial mechanisms, the system internalizes its own development effects. This property, which is fairly normal in the case of formal corporate institutions, is quite remarkable for an informal structure. Thus a spatial phenomenon contributes to a cumulative process of techno-economic development (Malecki, 1984: Lambooy, 1985).

Last but not least, we come finally to the question of behavioural changes associated with cultural dynamics. Rothwell and Wissema (1986) have analysed the place of cultural factors in technological change. We should like here to emphasize the fact that territorial "milieux" act as breeding places of cultural phenomena. Historians like Mumford (1961) have shown the leading role that "cities", which are a remarkable and powerful type of territorial organization, have played in the birth and growing up of western "civilisation". Andersson (1985) has explained the "fundamental role of metropolitan regions in the creative process" in terms of social synergies. He has "demonstrated that spatial structures play an important part in the creative process" and that "a very limited set of policy conclusions can be drawn from non-spatial theories of R&D" (1985: 5).

In our study of Sophia Antipolis, we have noticed the emergence of a specific cultural process: namely a general propensity both to generate new productive problems and to solve them collectively through rapid exchanges of ideas and informal co-operation (Perrin, 1986d). Formulae which were gathered during interviews are of interest: "un frottement technologique qui provoque l'imagination de solutions nouvelles" ... "un réseau de contacts d'où fourmillent des idées et des interrogations qui n'auraient pas eu lieu sans cela" ... "une émulation" ... "une excitation intellectuelle". This type of cultural phenomenon develops from active interchanges within a population of researchers, engineers and managers who practise technological innovation.

A further stage is reached when the emerging culture is recognized and socially valued. It then becomes a powerful vector for changing traditional behaviour and mentalities and for diffusing the propensity to innovate.

We may quote a statement by Andersson (1986: 8) which sounds like a theorem on local synergy: "if the social, psychological, geographical distance decreases between any two corporations of an economic system, then the balance growth of all corporations will increase". We have tried to demonstrate that territorial organizations decrease not only geographical distances but also socio-cultural ones. Finally, all these reasons explain why, as Gaffard (1986: 25) writes "le système territorial entre dans le fait productif et contribue à la création de technologie".

After this somewhat arid effort to give to the concept of local synergy a more rigorous content, it may be refreshing to recall that the impact of territorial structures on innovation has been superbly illustrated by historians like Fernand Braudel in "Civilisation Matérielle, Economie et Capitalisme: XVème au XVIIIème siècle", as well as by geographers like Pred in "The Spatial Dynamics of Urban Industrial Growth" and economists like Jacobs in "The Economy of Cities". Nevertheless, we will pursue our systematic approach of the local synergy concept in considering how useful it may be for regional technology policies in European countries and in which directions it should be developed.

REGIONAL TECHNOLOGY POLICIES AND THE DEVELOPMENT OF LOCAL SYNERGIES

We notice, first, that the concept of synergy may give a fruitful insight into the general meaning of public economic action in the field of development: namely to promote creative communications between agents. When conducted at a regional level, such a policy helps to increase local synergies. However this may be, we are not interested in the effect but in the process.

In what way can an approach in terms of local synergy be utilized for enhancing autonomous territorial capacities in the field of technological innovation?

In the context of recent French experience of decentralized regional planning (1984-8), we observe two main issues: technology poles and technology transfer projects. The first of these refers to local synergies of the incubator type. The second is concerned with the establishment of synergetic networks between research laboratories and industrial units, mainly small and medium-sized firms, within the whole region.

In recent articles, Stohr (1986a: 1986b) has proposed criteria for evaluating the success of science parks and other technological complexes, and grids for analysing the composition of all the types which have been recorded.

Our aim here is to go further into the question of how to implement regional technology policy, following a local synergy approach.

We shall first consider the diagnosis of territorial capacities in terms of advantages and deficiencies. We propose to analyze resources in terms of "know-how"; thus we define the concept of "collective technological innovation know-how" and we identify territorialised forms of that resource, in which local synerties are active. We shall call them **"innovation know-how concentrations"** ("gisements de savoir faire") and put forward a typology of these areas as a tool of analysis.

For elaborating a regional strategy, the second aim of this section, it is necessary to have a working hypothesis about the nature and direction of current technological evolution. We propose a paradigm which is derived from the concept of "technology clusters" ("grappes de technologies": GEST, 1985: Loinger and Peyrache, 1986). Regional policy may be conceived as the promotion of those functional synergies between identified resources which conform to the paradigm of technology dynamics. We differentiate levels or types of objectives which depend on the resources. Since these are localized it is possible to incorporate territorial effects - the local synergy effects - into the policy. To do this, it is also necessary to take into account the spatial characteristics of the communication structure of the region.

The final part of this section discusses examples which help to test the feasibility and the capacity of such an approach.

Identification of resources

As previously mentioned, innovative collective capacities are based upon connections between complementary know-how (from advanced research to production and marketing): in turn, connections are the result of organisational structures. We are interested in the territorial forms of such structures, which build upon local synergies.

From the point of view of regional development policy, the different types of know-how involved in these structures may be considered as resources which can be mobilized for new ventures either through labour market procedures, formal agreements or informal co-operation. Regional authorities need to identify spatial concentrations of such resources in terms of innovation know-how, and to evaluate their advantages and their deficiences.

We propose a typology which has been largely inspired by Loinger and Peyrache's recent work (1986) on "Le concept de grappe de technologies appliqué à l'espace économique régional". We distinguish four main types of "innovation know-how concentrations".

The first of these are the territorial incubation structures discussed earlier. We should however add that in order to constitute reliable resource concentrations, these need to be reasonably large and important. In Europe, this is certainly the case with the Cambridge area or West Berlin, for instance.

The other constraint is that research institutions need to be operationally linked to industry. In the case of Berlin, specific structures like B.I.G. or T.I.P. have been created with the participation of the Senate of the Land, in order to develop that connection. But, in most cases, especially in France, much still needs to be done to achieve significant thresholds (Perrin, 1986a).

The second type of resource area is where innovation know-how is locally generated through the new technology strategies of large companies.

As we have already noticed, traditional corporate policies inherited from the growth period which still often predominate in old multinationals, tend to localize research activities, production units and marketing services in different places. Thus they do not contribute to the formation of a territorialized resource. It is only when, for productivity purposes, these companies develop regional sub-contracting policies involving technology transfer that local know-how is created which can be utilized by regional policy.

In contrast to such traditional policies, local innovative capacities may emerge from technology strategies which have only recently been adopted, mainly by younger multinationals. In Loinger and Peyrache's phrase (1986: 7), they operate under a "logique de déconcentration par segment technologique homogène territorialisé".

Research and development, production, and marketing units operate in terms of very specific and rapidly changing markets in a kind of continuous circular interacting process. For these firms, proximity is a factor of synergy. Decentralized teams are given complete freedom to develop synergetic relations with other neighbouring academic laboratories and R&D enterprises. "Le jeu", Loinger and Peyrache (1986: 9) write "consistera pour le directeur de l'établissement - quasi chef d'entreprise d'une grosse PME -à s'enraciner au maximum dans le tissu local, avec deux aspects dominants: la recherche scientifique en prenant appui sur tel laboratoire public de recherche; la sous-traitance, en mettant en competition les entreprises de la zone géographique d'implantation, quitte à diffuser au maximum les informations nécessaires". Thus the innovation know-how which was originally internalized by the private institution later on diffuses through trans-institutional connections and labour market channels within the area.

The third type of territorialized innovation know-how also arises from the activities of large companies for which the logic of local synergy prevails over that of spatial disjunction, but in a different way. The firms involved here have operated for many years in their region of birth, and have dominated its industrial development and structure. They are also distinctive in terms of the nature of their activity: they all manufacture complex products. Good examples include Fiat and Olivetti in the Turin-Piedmont area, Peugeot in Franche Comté, Philips in the Southern Netherlands, and Volkswagen in Hanover. These firms have concentrated in their core region their most important R&D departments, production units, and their headquarters. In general, they organize around them networks of sub-contractors and specific equipment manufacturers especially for R&D activities. With these other firms, they develop circular technology transfers. Quite often, independent specialized machinery activities also grow up in the seed bed environment afforded by the region.

For companies whose decision-making headquarters are located somewhere else, the situation must be viewed differently since this means that they operate along spatial disjunction lines. Thus their R&D units and their sub-contractors also have different locations. This is the case, for example, with the shipbuilding industries of Newcastle (Amin, 1986) and the Toulon area.

As for firms which operate in processing sectors (iron

and steel, for instance) and which have also remained concentrated in their traditional location zone, the problem is again different. Here it is the nature of the activity which is not favourable to the diffusion of innovation capacities. Nevertheless, in old industrial regions, technical universities or engineering schools have often existed for a long time. Even if leading companies have not been successful in restructuring their activities, academic institutions have frequently progressively promoted new sectors and indeed new technologies. They have even sometimes developed joint research programmes in these fields in close co-operation with industry. They may therefore constitute a strategic resource for regional technology policies.

The fourth and last type of regional concentration of technical know-how is associated with sets of SMEs having similar industrial profiles. Examples may be found in Prato, Cholet, Mulhouse, and Roubaix for ready-to-wear, shoes, clothing, and textile activities, and in the Swiss Jura for watch-making.

The organizational structure of these areas is the result of an historical process in which economic links have been reinforced by social traditions. Incubation processes have taken place in urban systems which are tightly connected with these activities or they have evolved in the context of traditional networks of family relations. This type of substratum is ambivalent. Sometimes it may have counter-effects. As Loinger and Peyrache (1986: 13) point out: "Les méthodes et les savoir faire ne sont jamais que l'expression de formes de rapports sociaux de production" ... "Pour ne pas modifier les rôles et les hierarchies, une certaine propension au maintien du statu quo en matière de méthodes de production peut entrainer une dégénerescence du milieu professionnel".

In some cases, commercial structures have helped to isolate these production areas from external world and market evolutions, as was the case with the watch-making zone of the Swiss Jura. The opposite is true for the Italian region called by Piore and Sabel (1983) "Third Italy", where long-established social features exist such as extended families and family enterprises. "It is a vast network of very small enterprises spread through villages and small cities of central and northeast Italy ... and ranging across a wide spectrum of sectors from shoes, ceramics, textiles and garments on one side to motorcycles, agricultural equipment, automotive parts and machine tools, on the other "

(Piore and Sabel, 1983: 392). Stohr (1986a: 25) stresses the existence of a highly innovative feedback mechanism: "within firms, by close co-operation between owners, designers, technicians and production workers in which hierarchical distinctions tend to be treated as formalities; between firms by intensive exchanges of ideas between owners, skilled workers and small consulting firms, as well as direct collaboration between dynamic small firms which share the cost of innovation, exchange orders mutually, have joint marketing, technical services ...Where invention creates demand and invention is also collective, collaboration is a natural result".

As far as technological dynamism is concerned, we must emphasize two properties of these structures. The first is a considerable capacity for product development. As a matter of fact, it has been observed that this very important phase of the innovative process is too often shortcut by SMEs. The organizations mentioned earlier also have considerable experience in connecting markets and changes in demand to new opportunities offered by advances in technology. Secondly, through time, machine building and equipment activities have developed to serve the local market provided by the dominant industry. In due course, accumulated expertise has bred specialisation processes in related machine-tools production which have in turn resulted in further diversification towards other sectors of demand. Finally, thanks to these new resources and since close links have been maintained between production and equipment, such areas exhibit a considerable capacity for promoting and incorporating technological progress.

We have just described four types of territorialized industrial clusters in which innovation know-how conducive to the development of new technologies may accumulate. Each of these proceeds from a specific generic organization in which the institutional component plays a major role.

We might add another category which combines elements of the preceding ones and which is developing within highly inter-related new technology sectors. The agglomeration of advanced research institutions (but not incubators), of large production units of multinationals (but not as part of any deliberate territorial strategy), and of subcontracting and service capacities, combined with the development of international transport facilities, appears to constitute a seed-bed for generating new activities through local spin-off effects, and for attracting small and medium-sized firms. The phenomenon is observable in the last few years, for instance, within the Marseille metropolitan area, around Aix-en-Provence, in the sector of applied electronics. A pool of high-technology know-how and information networks are emerging which may in due course constitute innovation resources.

In helping to identify regional potentials, the typology outlined above constitutes a tool which may help to elaborate technology strategies.

Elaboration of strategies

The objectives of any technology policy must be related to the leading evolutionary tendencies in that field. How, in the present decade, can technology dynamics be characterized? We shall propose an hypothesis which is derived from the concept of a "technology cluster". Following this, we shall define different levels or types of objectives, depending on the composition of regional resources identified in terms of "high technology innovative know-how concentrations". Solutions are conceived as "synergetic paths". The corresponding strategies consist of making use of territorial structuring capabilities (the local synergy approach). To do this, a spatial analysis of the communication externalities of the regional socio-economic system is necessary.

The concept of "technology clusters" ("grappes de technologies") has been proposed by a team under the collective name of GEST (1985). It is built from case studies of the technology strategies of different multinationals. The concept is a "tool for analysing an evolution" (GEST, 1985: 19). At the level of the firm, "les grappes de technologies sont le produit de la perception, à un moment donné, des synergies possibles en termes technico-économiques qui ouvrent un champ cohèrent de développement pour ses activités" (GEST, 1985: 18). But the authors emphasize the fact that the approach may be applied at a macro-sectorial level for analysing the present transformation of the production system. "Elle permet de dégager les principales tendances d'évolution du système industriel" (GEST, 1985: 19). Loinger and Peyrache (1986) have extended the analysis to spatial regional problems.

The concept of a technology cluster combines three ideas. These are the articulation of fundamental and applied research following generic technologies, their organization in varied combinations corresponding to new products which are developed for different sectors and markets, and the capitalisation of know-how that can, later on, be mobilized and applied in responding to unpredictable market evolutions. In other words, "le fondement de la stratégie de grappe est la maitrise d'un savoir faire combinant un ensemble architecturé de technologies et sa valorisation sur un espace de produits et de marchés le plus vaste possible" (GEST, 1985: 49).

From this research, it may be concluded that nowadays, technological progress proceeds from combinations of generic technologies, from their close articulation with product development, and from the openness of these combinatorial and development processes in terms of sectors and markets.

Depending on regional resources, two main levels and types of objectives may be defined. One is the promotion of open combinations of generic technologies and of associated product developments. Advanced regions which have enough resources for implementing those combinations, could, thus, become places where new technology clusters are generated. The other is the promotion of at least one generic technology and related product developments.

This distinction gives the frame of a grid which will be enriched through case studies.

Both types of policies are conceived as **synergetic paths.** This means that, in order to achieve their goals, regional actors have, firstly, to stimulate constructive connections between the innovation resources that have been identified with their strengths and weaknesses and with their territorial and organizational contents, and, secondly, to make the best use of local synergies.

"Territorial resource concentrations" have their own potentials which may be exploited and fostered: for instance, in implanting complementary factors which, in turn, will benefit from the existence of local networks. But the main target of the policy is the connection of resources.

In this case, territorial synergies depend on a group of interrelated factors. These include the spatial urban-industrial structure and transport system of the region, intercity mobility practices (thus, for instance, within the Rimcity of Netherlands there is no problem of accessibility, even between distant points), and inter-institutional communication behaviour, particularly between public and private structures, large and small companies, and sectorial and territorial administrations. Great differences exist between centralised and decentralised systems, and in the latter, between the capital region - such as the Parisian agglomeration in France - and the rest of the country.

This approach and these tools may be tested by applying them to the many European case-studies which have recently been undertaken - especially under the initiative of the French-speaking section of the Regional Science Association (ASRDLF) and of the GREMI (GREMI, 1986a: b: Federwisch and Zoller, 1986) - on new technology and territorial development. Do they bring more coherent and deeper insights into regional technology problems and associated policies? The final section of this chapter briefly assesses this issue.

ILLUSTRATIONS

Advanced regions

These are areas in which efficient innovation areas of different sorts co-exist and where high technology industries are diversified so that policies directed towards generic combination could be initiated. Nevertheless, communication barriers are sometimes so strong that preliminary policy measures are necessary.

This seems to be the case with the Paris Sud area (Decoster and Tabariés, 1986). This contains the largest single accumulation and concentration in France of private and public research institutions, prestigious "grandes Ecoles", and leading large and small high-technology companies. Despite this, however, synergies and cross fertilisation are minimal, mainly for communication reasons. Firstly, in this capital region of the historically centralized French system, sectorialisation and inter-institutional barriers are particularly strong. Each unit behaves like a citadel jealous of its power and of its supposed superiority. "Recherche et industrie se regardent en chiens de faience et se reprochent mutuellement leur peu d'envergure" (Chambon and Dyan, 1985: 145). This cultural phenomenon has also been noted in Japan's major Science City at Tsukuba. Secondly, the geographical scale and physical design of the area, which is very close to the inner city, are not appropriate for the development of local synergies. The whole phenomenon is chiefly the result of a simple deconcentration policy undertaken for physical planning reasons by "the District Urbain de Paris". It has nothing to do with the concept of Technopolis.

A contrast with this example is provided by the Milan Technological Centre project (Camagni, 1986) located in the eastern part of the city, which is developing rapidly under public and private initiatives.

West Berlin is another advanced, diversified industrial area highly endowed with research centres and technological universities. But for the same type of historical reasons there have been barriers to communication between universities and industries in the city, as well as between large and small enterprises.

In recent years, however, a synergy-oriented policy has been initiated by academic, administrative and private actors. It is focussed on technology transfers between research institutions and SMEs, and is especially concerned with helping the latter during the difficult development phase. Two structures of the incubator-nursery type have been successively implemented. The "B.I.G." project brings together university laboratories, small high-technology firms and nursery services. The "T.V.A" (Technologie Vermittlungs Agentur) scheme is conceived as a network covering the whole regional area and extending outside in order to create "an international innovation network". It is animated and controlled by an independent agency whose financing comes from both private enterprises and state administrations.

The Rhône-Alpes region provides an interesting example of a technology policy of the combinatorial type. It is a highly developed urban-industrial area with a strongly established tradition of autonomy characterized by internal communication of the decentralized type. All four categories of technological resource concentrations are identifiable here: sometimes in the same locality, as for instance in the Grenoble area where segments of electronic generic technology are developing in a "cluster" fashion (De Bernardy and Boisgontier, 1986: Loinger, 1986).

Two types of "combinatorial" policies are currently being implemented in this region. They involve two different forms of local synergy arrangement. One is a network of science parks in the Lyon Metropolitan Area (Tesse, 1985); the other is the so-called "pôle productique régional" (Bellet and Boureille, 1985: 1986).

What is the synergetic logic of the first project? Firstly, in order to develop communications between research institutions and industry and to extend them to SMEs, three incubation and nursery structures have been planned. Secondly, two of these nursery schemes are involved in generic technologies: the Gerland area is devoted to bio-technology, the western park to electronics. The third zone is open to other activities in which the region possesses special expertise. Thirdly, functional synergies between the three poles may be expected since they are localised in the same urban area which is a very powerful communication heartland.

The second project is concerned with the field of automated-production systems, which in itself combines generic technologies such as electronics, new materials and energy sources with more specific ones such as micro-mechanics and digital control. The objective of the project is to develop a capacity for designing and building automated systems adapted to the main regional sectors of activity, especially those traditional industries which are confronted with the need to restructure their operations. The eventual aim is to export skills and technologies that will have been developed on the basis of serving local markets. The organizational structure of the pole combines functional and local synergies through a segmentation on three levels (Bellet and Ribeille, 1986). One refers to the development of basic components for automated systems. The corresponding agents are located in production centres such as Lyon, Grenoble, St Etienne and Valence. Another is concerned with the adaptation of robots to the specific needs of particular regional industrial activities. The places which have been chosen are Roannes for textile machinery, Oyonnax for plastics processing, and Cluses for screw-cutting production. The third level is devoted to technology transfers. It is localized in those places where R&D on robots is developing. Thus the "pôle productique" is organized as a regional network that connects all the actors - centres, agencies, enterprises - who have joined the project and who are scattered throughout the whole territory. The network is controlled by an independent association sponsored by all the constituent members.

Other regions

We subsume under this category all those regions whose long-term objective could be to promote local industrial capacities in at least one generic technology and in the development of associated products. These regions can also be differentiated by the way they combine two aforementioned criteria: the resource composition, in terms of "innovation know-how concentrations" (the four categories defined earlier) and the "communication profile" characteristics.

The present discussion will however be restricted to examples of the problematic and constantly-debated question of old industrial regions such as Newcastle (Amin, 1986), Charleroi (Alaluf, 1986), Besançon (Pottier and Touati, 1986), the Swiss Jura (Maillat, 1986), Liguria (Camagni, 1986) and Nord Pas de Calais (Schoenberger, 1984).

Generally, resource concentrations of this kind are easily identifiable, since they have developed from geographical and historical endowments and are based on industries such as iron and steel, textiles, ship-building and micro mechanics.

In areas where industry is dominated by externally controlled multinationals, innovation resources are generally no longer present (Newcastle, North East Lorraine, La Seyne-La Ciotat...). An important exception to this in some cases is the existence of engineering schools with long traditions. In other areas, where the industrial structure is characterized by sets of SMEs or by strongly-rooted large companies, a focus of redeployment may be constituted by induced equipment goods and machine tools activities as well as by sub-contractors with good technical know-how, as discussed above. Both types possess the capacity for diversification. But to achieve this, their regeneration must incorporate external generic technologies; and they have to learn how to develop new products, since they have previously been accustomed to serving captive markets.

Local synergy policies for older industrial regions would favour solutions in which R&D institutions and technological high-schools were implanted in the area and networks of technology transfer set up with a strong emphasis on product development. This could be done, for instance, through specific contracts (contracts for the development of new products, between research agencies and private firms) sponsored by regional authorities, as is the case in Japanese technopolises.

The synergetic web may also be extended: downwards, to traditional production units (Schoenberger, 1984) by directing the modernisation of the machine sector in tune with their specific needs, and upwards, to scientific public laboratories, by orienting the operations of product development towards related and very specific specialisations.

Spatial synergies should also be encouraged by polarizing around these new capacities complementary structures such as nursery agencies, export services and advanced training institutions, and by attracting establishments of multinationals as suggested in Pottier's (1986) analysis of Besançon.

FINAL REMARKS

We have explored the concept of local synergy in relation to technological innovation. We have tried to give it some theoretical consistency and to derive useful approaches for regional policy. But these are only the first steps. Much still remains to be done. It would be interesting, for instance, to identify the networks associated with the four types of resource concentration areas that we have distinguished and to compare their structures (and their articulations with the outside world) in order to make these tools of analysis more operational. The resistance of traditional behaviour patterns and networks to innovative strategies and organizations is such that the connecting structures that have been described would have to be reinforced by institutional means, especially with regard to the mobility of agents between public and private establishments and the development of collaborative research between academic institutions and enterprises through contractual procedures. We also know little about the specific dynamics of the development of collective innovation know-how: learning processes, training procedures, labour market mechanisms...

But, as a first step, the applications that we have just outlined could be extended to more case studies.

REFERENCES

Alaluf, M. (1986) Situation économique, facteurs de redéploiement et innovation technologique; le cas de Charleroi. In P. Aydalot (ed.), Milieux innovateurs en Europe, GREMI, Paris, pp. 163-93

Amin, A. and Thwaites, A. (1986) Technical change and the local economy: the case of the Northern Region (UK). In P. Aydalot (ed.), Milieux innovateurs en Europe,

GREMI, Paris, pp. 129-61
Andersson, A. (1985) Creativity and regional development, Papers of the Regional Science Association 56, pp. 5-20
Andersson, A. (1986) Creativity, complexity and economic development. Paper for the Conference on Innovation diffusion, 17-21 March, Venice
Aydalot, P. (1986) Trajectoires technologiques et milieux innovateurs. In P. Aydalot (ed.), Milieux innovateurs en Europe, GREMI, Paris, pp. 345-61
Bellet, M. and Boureille, B. (1985) Conditions du développement local à partir d'un pôle productique: le cas de la région stéphanoise, Revue d'Economie Régionale et Urbaine, 4, pp. 741-54
Bellet, M. and Boureille, B. (1986) Pôles productiques régionaux ou trajectoires territoriales d'un nouveau paradigm technologique. In Association de Science Régionale de Langue Française, Technologies nouvelles et développement régional, GREMI, Paris, pp. 29-35
Camagni, R. (1986) Innovation and territory: the Milan high-tech and innovation field. In P. Aydalot (ed.), Milieux innovateurs en Europe, GREMI, Paris, pp. 101-25
Camagni, R. (1986) Robotique industrielle et revitalisation du Nord Ouest italien. In J. Federwisch and H. Zoller (eds) Technologie nouvelle et ruptures régionales, Economica, Paris, pp. 59-81
Catin, M. (1985) Effets externes, Economica, Paris
Chambon, P. and Dyan, B. (1985) La mase critique ou le concept de Paris Sud, Autrement, 74, November, pp. 143-8
De Bernardy, M. and Boisgontier, P. (1986) Les micro entrerprises de la région grenobloise et leurs relations au tissu local, Revue d'Economie Régionale et Urbaine, 5, pp. 645-63
Decoster, E. and Tabariés, M. (1986) L'innovation dans un pôle scientifique et technologique; le cas de la Cité Scientifique Ile de France Sud. In P. Aydalot (ed.), Milieux innovateurs en Europe, GREMI, Paris, pp. 79-100
Federwisch, J. and Zoller, H. (eds) (1986) Technologie nouvelle et ruptures régionales, Economica, Paris
Gaffard, J. L. (1986) Restructuration de l'espace economique et trajectoires économiques. In P. Aydalot (ed.), Milieux innovateurs en Europe, GREMI, Paris, pp. 17-27
GEST (1985) Grappes technologiques et stratégies

industrielles, Ministère de la Recherche et de l'Industrie, Etude no 57, Paris

GREMI (1986a) Milieux innovateurs en Europe. P. Aydalot (ed.), Paris

GREMI (1986b) Technologies nouvelles et développement régional, Association de Science Régionale de Langue Française (ed.), Paris

Lambooy, J.G. and de Jong, M.W. (1986) Urban dynamics and the new firm: the position of Amsterdam in the Northern Rimcity. In D. Keeble and E. Wever (eds), New firms and regional development in Europe, Croom Helm, London, pp. 203-23

Loinger, G. and Peyrache, V. (1986) Le concept de grappes de technologies appliqué à l'espace economique régional. In Association de Science Régionale de Langue Française, Technologies nouvelles et développement régional, GREMI, Paris, pp. 461-71

Maillat, D. (1986) Les milieux innovateurs; le cas de l'arc jurassien suisse. In P. Aydalot (ed.), Milieux innovateurs en Europe, GREMI, Paris, pp. 217-46

Malecki, E.J. (1984) High technology and local economic development, Journal of the American Planners Association, 4, pp. 262-9

Mariotti, S. and Ricotta, E. (1986) Diversification agreements among firms and innovative behaviour. Paper for the Conference on Innovation diffusion, March 14-21, Venice

Mumford, L. (1960) La cité à travers l'histoire, Seuil, Paris

Pavitt, K. (1984) Sectoral patterns of technological change: towards a taxonomy and a theory, Research Policy, 13, pp. 343-75

Perrin, J.C. (1974) Le développement régional, PUF, Paris

Perrin, J.C. (1986a) Le phénomène Sophia Antipolis dans son environnement régional. In P. Aydalot (ed.), Les milieux innovateurs en Europe, GREMI, Paris, pp. 283-303

Perrin, J.C. (1986b) Planification décentralisée, mutations technologiques, PMI - Propositions pour une politique industrielle régionale. Rapport pour le Commissariat Général du Plan, Mimeo. CER, Aix-en-Provence

Perrin, J.C. (1986c) Les PME de haute technologie à Valbonne Sophia Antipolis, Revue d'Economie Régionale et Urbaine, 5, pp. 629-45

Perrin, J. C. (1986d) Technopoles et développement. Rapport pour le CNRS, Mimeo. CER, Aix-en-Provence

Perrin, J.C. (1987) A deconcentrated technology policy-lessons from the Sophia Antipolis experience. Environment and Planning, special issue on Technology Policy

Piore, M. and Sabel, C. (1983) Italian small business development: lessons for U.S. industrial policy. In Zysman J. and Tyson L. and (eds), American industry in international competition: government policies and corporate strategies, Cornell University Press, Ithaca

Planque, B. (1983) Innovation et développement régional, Economica, Paris

Pottier, C. and Touati, P.Y. (1986) Les conditions de l'innovation dans les régions d'industrialisation ancienne. In P. Aydalot (ed.), Les milieux innovateurs en Europe, GREMI, Paris, pp. 247-66

Rispoli, M. and Volpato, G. (1986) Innovation acquisition - an unavoidable challenge to small firms. Paper for the Conference on Innovation diffusion, March 14-21, Venice

Rothwell, R. and Wissema, H. (1986) Technology, culture and public policy, Technovation, 4, pp. 91-115

Schoenberger, E. (1984) Développement régional et technologies de pointe dans le Nord-Pas-de-Calais, Revue d'Economie Régionale et Urbaine, 4, pp. 649-67

Stohr, W. (1986a) Territorial innovation complexes, IIR Discussion Paper 28, Vienna

Stohr, W. (1986b) The role of local/regional synergism for innovation. Association de Science Régionale de Langue Française, Technologies nouvelles et développement régional, GREMI, Paris, pp. 373-83

Tesse, P.Y. (1985) Du sexe des technopoles - Lyon. Autrement, 74, pp. 156-9

Chapter 8

Economic and Territorial Conditions for Indigenous Revival in Europe's Industrial Regions

Denis Maillat and Jean-Yves Vasserot

DEFINITION OF THE ISSUE

The connection between technological and economic development is today widely acknowledged. In many regions, the importance of this link has been underscored by its determining role over the past 15 years in structural and technological change. A new configuration of economic interactions has thus been emerging (Aydalot, 1986; Castells, 1985). Today's new and highly innovative centres of economic activity are firmly stamped with the seal of new technologies, particularly those in the high-technology category (Route 128, Silicon Valley, etc.). These new vanguard industries, including electronics, telecommunications, industrial materials, biotechnology, artificial intelligence, fibre optics and robotics, have spearheaded the majority of current process and product innovations. They have also triggered a resurgence of interest in the entrepreneurial spirit (Lambooy, 1984; Martin, 1986; Maillat, Schoepfer and Voillat, 1984; Dieperinck and Nijkamp, 1985). Their timely emergence is likely greatly to facilitate and stimulate the remodelling of the production systems of industrial regions. The following study constitutes an analysis of the conditions required for this.

The degree of success recorded by regions geared towards new technologies has served increasingly to underscore the stagnation, immobility and lack of adaptability of industrial regions. In fact, industrial regions often appear comparatively as the least innovative of all regions.

Moreover, industrial regions fail to offer sufficiently favourable conditions to attract high-technology industries (Camagni and Rabellotti, 1986; Nishioka, 1983). Their revival must therefore be considered as dependent on their ability to assimilate and use new technologies rather than in terms of an influx of new enterprises (Maillat, 1984). This ability hinges on the propensity for change exhibited by activities and firms already established in the region (territorialized production systems) as well as on the availability of opportunities to create and develop new enterprises (Daynac, Millien and Cunat, 1983; Lambooy, 1984; Maillat, Schoepfer and Voillat, 1984).

The current technological revolution has underscored the fact that techno-scientific progress tends to exacerbate the obsolescence of traditional structures (Perrin, 1984). Territorial production systems, like enterprises, face the need to adapt. Successful transformation has outstripped growth as today's major concern. This transformation can be facilitated by new technologies which "offer the means to revive traditional industries, create new industries and restructure and diversify economic activity" (OCDE, 1984). The multiple combinative possibilities of new with traditional technologies offer a range of opportunities ripe for seizure by a number of industrial regions. This means that these regions can recover, under certain conditions, their innovative and creative capacity.

According to Andersson (1985), creativity is a social phenomenon which occurs primarily in regions characterised by high skill levels (knowledge embodied in the population and activities of a region), a variety of university and cultural activities, well-developed internal and external communications systems, widespread consciousness of unmet needs (hence openness towards change) and structural instability (which, owing to the uncertainty it generates, tends to promote interaction between different spheres of regional skills and thus trigger change).

This definition of creative regions does not mesh exactly with the profile of industrial regions. However, many industrial regions display several of the factors cited by Andersson, particularly knowledge stored in the form of know-how. This know-how encompasses the practical and intellectual mastery of all the technical skills involved in a production system. Studies in different European regions have shown that this type of know-how can be successfully adapted to new technologies. This has been the case, for example, in the cradle of the Swiss watch-making industry (Maillat, Schoepfer and Voillat, 1984; Maillat, 1985) as well as in the Franche-Comté, Alsace, Rhône-Alpes (Pottier, 1984), St Etienne (Peyrache, 1986; Thomas, 1986) and Nord-

Pas-de-Calais (Cunat, 1986), regions under study here. The territorial systems of these regions harbour resources rife with realisable economic potential (Perrin, 1983). Their specialised industrial profiles constitute assets utilisable in the reindustrialisation process (Daynac, Millien and Cunat, 1984; Pottier, 1984). A typical example is initial specialisation in machine tools which facilitates transition to the manufacturing of numerical control machines and robots.

Moreover, the need for revival is particularly acute within industrial regions because of the considerable job losses they have suffered. Structural instability has left them little choice but to react or succumb to decay. The path towards revival has been paved for them by the advent of new technologies offering increased scope for adaptation and conversion. Although the direction which restructuring takes tends to be shaped by traditional industrial profiles, its initiation is not necessarily voluntary. In the majority of the above-mentioned industrial regions, impetus for change stemmed from a blow dealt from the outside (such as the emergence of new technologies or the impact of international competition) which caused an upheaval throwing the region into a state of uncertainty and instability conducive to restructuring.

Undoubtedly, restructuring is not always easy. To achieve it, industrial regions must draw on their deepest resources. They must not only mobilize endogenous potential, but also absorb new activities (Maillat, 1984; Pottier, 1984). In mobilising their potential and integrating outside enterprises, they face a number of obstacles. Tradition as well as the complacency of a handful of privileged regional firms often stand in the way of change. In addition, industry today no longer constitutes the sole driving force behind production systems. Economic development is increasingly becoming a function of the growing interconnections between secondary and tertiary activities (Bailly and Maillat, 1986; Lambooy and Tordoir, 1985; Malecki, 1983). Industrial region production systems must consequently include service activities both upstream and downstream, from the research to the marketing stages. Restructuring must operate at two levels: that of industrial activities and that of producer services. In other words, successful restructuring depends upon the ability of industrial regions to re-establish coherence within territorial production systems (Perrin, 1984; Maillat, 1984). The need to consider the territorial aspect of production systems is obvious in the case of industrial regions since "the production process must henceforth be viewed as a changing reflection of the environment" (Gaffard, 1986). On the one hand, the regional environment must be changed. Yet on the other hand, it must contain the seeds of change. Following this line of thinking, "the key aspect of localization is the creation, within a specific geographical area, of institutions designed to promote co-operation between enterprises and establishments grafting themselves onto traditional market relations. ... Proximity is an important component of the ability to exchange information and therefore create new technology" (Gaffard, 1986).

The ultimate goal of restructuring for industrial regions is of course to recover comparative advantages capable of giving them a renewed competitive edge and innovative role. They must move into new functional production spheres. At the same time, they must modify the territorial aspect of their production systems in an effort to become, on the strength of their accumulated and adaptable know-how, co-producers of, rather than mere receptacles for, technology and innovation (Perrin, 1984).

The present study focusses primarily on the functioning of territorial production systems. This hinges on three groups of elements which play a crucial role in revival, namely:

1. The territorial production network (nature of enterprises and of their regional insertion, coherence of the entire range of activities, etc.).
2. The functioning of the labour market (types of occupations, types of mobility channels).
3. The elements comprising the scientific network (training systems, research institutes, regional production and accumulation of knowledge).

The hypothesis which will be considered here is that ties connecting the above-mentioned three groups of elements constitute networks which carry, disseminate and develop the knowledge necessary for revival.

THE PRODUCTION NETWORK

Technological change and innovation

The concepts of technological change and innovation must

be precisely defined before an analysis can be made of the production systems of traditional industrial regions and of the characteristics of firms capable of technological change. Although there is currently a general tendency to focus on high-technology industries and therefore to channel research into new technologies, differences exist between the problems associated with the development of traditional industrial regions and those of regions already abounding in advanced-technology sectors. This is underscored by the findings of numerous studies (Oakey, Thwaites and Nash, 1980; Aydalot, 1984; Hall, 1985; Meyer-Krahmer, 1985) tending to show that traditional industrial regions, which are often economically depressed areas, have rates of innovation that are comparatively lower than other regions.

A distinction must be drawn between innovations which consist of the improvement of products and processes (incremental innovations) and radical innovations which trigger revolutionary changes liable to modify "techno-economic paradigms" (Camagni and Rabellotti, 1986). The first case refers to a regular updating and adapting of existing products and processes in response to the development of technology. This kind of innovation is a continuous process, known as technological change, rather than an innovative process in the strict sense of the term.

Such technological change, although less significant than radical innovation, is crucial to the future of the enterprises concerned. It triggers product alterations enabling enterprises to maintain and expand their market shares. It constitutes proof that they are capable of assimilating and using new technologies. Thus, not only major or basic innovations are important to the maintenance of technological progress within regional industries, but also technological updating and adapting of the products already being manufactured (Amin and Thwaites, 1986). It is this renovation process which enables traditional industries to develop future prospects. In the Swiss Jura Arc region, for example, three-quarters of all product innovations can be traced to improvements in existing products, while only the remaining quarter results half from diversification and half from conversion to high-technology products (Maillat and Vasserot, 1986).

Innovative enterprises and their regional ties

The regional distribution of product and process innovations depends to a large extent upon the characteristics of the enterprises in question, particularly upon their ability to carry out technological change (Boulianne and Maillat, 1984; Malecki, 1983; Meyer-Krahmer, 1985). Certain enterprises are more capable of innovation than others. This depends upon their internal organization, the types of products they manufacture, the nature of their labourforces, their store of knowledge, their development policies and resources, and other factors. The profile of an innovative enterprise can thus be approximately drawn. In many respects, the key factor is the location of development and innovation policy functions. Innovations within capital-intensive industries requiring high R & D investment generally stem from large enterprises. Small and medium-sized enterprises (SMEs) tend to innovate in areas where capital investment and development costs remain low and barriers to new enterprises weak (Rothwell and Zegveld, 1983).

A CURDS study of the Northern Region (Amin and Thwaites, 1986) points in the same direction by demonstrating the importance of enterprise structure (group, independent establishment, subsidiary, etc.) and capital ownership (indigenous or external to the region) in determining the technological path of a firm. The study's examination of the capital structure of Northern Region enterprises revealed that three-quarters of locally-established firms consisted of parent or subsidiary establishments belonging to groups or concerns with headquarters and decision-making centres located beyond regional or even national borders. During the period of recession experienced by the region, this category of enterprise was alone responsible for 80% of the overall local loss of jobs. This shows the degree of domination exerted by these large group establishments operating within declining sectors which have drastically reduced their activity in the Northern Region without helping to set in motion regional conversion or diversification processes. Although there are few SMEs in this region, it is less the large size of its enterprises than the combined effect of their size, structure and capital ownership which has played a negative role resulting in a regional concentration of low-level production activities and an inability to incorporate functions generative of technological change.

This example clearly demonstrates the importance of firm regional ties, that is the territorialization of functions conducive to technological development. "R & D activity and personnel represent the core of what we understand to be the capacity for positive regional change, along with the question of economic power and control" (Malecki, 1983). Thus, regions must be able to escape the logic of the filtering-down process if enterprises are to enjoy conditions favourable to innovation. In this respect, SMEs have a determining role to play.

Outward-orientedness and the role of SMEs

Large establishments and large companies cease to grow and expand in periods of technological change, and numerous observations point to the fact that SMEs, owing to their greater flexibility, are in a better position to face uncertainty.

Innovative SMEs (defined by their R & D activity) show above-average employment growth rates, even in declining sectors (Meyer-Krahmer, 1985). It is thus important to achieve a better understanding of what distinguishes them from other firms. Meyer-Krahmer's study (1985) of innovative SMEs in the Federal Republic of Germany is particularly eloquent and thorough in this respect. The most innovative SMEs were found to engage in regular R & D activity, manufacture several types of products (avoiding dependence on a single product) and rely heavily on exporting (the bulk of their sales were made to foreign markets). The innovative nature of an enterprise thus seems to be closely linked to a three-fold focus on R & D, product diversification and export marketing. Enterprises which do little or no innovating are characterized by a lack of R & D activity and the tendency to manufacture only one type of product and market it locally. Even without a more precise explanation of the causality of this occurrence, it can be seen that one of the major features distinguishing innovative from non-innovative firms in terms of market area is the innovative firm's increasingly wide market outreach.

This observation is not without relevance for traditional industrial regions dominated either by mono-industrial structures or by large enterprises which shape industrial region profiles to suit their needs. Sub-contracting practised to the point of creating dependence, as in the St Etienne region (Peyrache, 1986), only exacerbates the problems of market diversification and, consequently, of market expansion.

This leads to consideration of the concept of the degree

of outward-orientedness of regional enterprises, another factor shaping the innovative behaviour of SMEs. A distinction can be drawn between high, medium and low degrees of outward-orientedness (Meyer-Krahmer, 1985). This distinction contributes to the analysis by determining the capacity of firms to absorb external sources of knowledge and know-how necessary to their technological development. Enterprises with high degrees of outward-orientedness invariably succeed, whatever their location, in obtaining the information they seek (through formal contacts with experts, consultants, research laboratories, etc.). This is the case with advanced-technology enterprises with location constraints which are independent from these information factors (footloose firms). Enterprises with medium degrees of outward-orientedness show a preference for solving their problems internally and have only occasional, informal contacts with outside providers of information. However, the local availability of sources of know-how, such as universities and technical schools, increases their reliance on external inputs. A good example of this kind of enterprise is provided by SMEs in the Swiss Jura Arc Region. These enterprises, which adapt and improve their products in response to changing technological conditions (Maillat and Vasserot, 1986) have benefitted from technological interface bodies which facilitate the acquisition and diffusion of technology (Maillat, Schoepfer and Voillat, 1984; Vasserot, 1986). As for enterprises with low degrees of outward-orientedness, they remain impervious to the local availability of information sources owing to obstacles of a basically internal nature, such as ferocious resistance to outside interference.

Network incorporation and producer services

In traditional industrial regions, enterprises do not act in isolation. They are generally connected to each other through entire or partial networks of activities. In fact, such networks have sometimes been used to define regions. The advantage of this concept is to provide regional production systems with internal coherence, especially in regions abounding in SMEs. This coherence stems from the organizational principles characteristic of activity networks. It lends meaning to the technological and economic effects of complementarity and interdependence. Thus, in the traditional industrial regions under analysis here, the ability to change hinges not on any particular form of specialization, but on a web of interdependent activities conducive to sub-contracting and partnership arrangements. From this viewpoint, activity networks indicate that "looking backwards from an end product (a watch, a car), it is not helpful to study only the 'watch industry' or the 'car industry' but it is useful to study a 'tree' of activities diagonally through sectors and branches" (Lambooy, 1984).

Furthermore, consideration must be given to the fact that national and international forces interact with regional forces. Integration of these various forces is ensured by the existence of complete and partial regional networks.

An example of this has been provided by the integration of the electronics network into the Swiss watch-making region (Maillat, 1984). Similar examples can be seen in regions dominated by machine-tool and textile industries. Where restructuring is concerned, the ability to incorporate a new network component (such as electronics) into a more traditional network (such as mechanical engineering) is essential.

This process undoubtedly often leads to job losses. However, it has a positive impact in terms of revival potential since it introduces new technologies into regional production systems.

This stimulates the development of new interdependences, and the emergence or resurgence of chain reactions, without destroying the traditional coherence of these systems. In this sense, a relationship exists between traditional and newly emerging systems. This relationship is reflected in particular by the integration of new and traditional know-how. Both new and traditional enterprises benefit from regional skills by being incorporated in regional networks.

Discovery of the interdependence between activities has gradually led to a broadening of the network concept beyond its purely industrial definition. Production systems have been increasingly moving towards the integration of manufacturing and service activities, particularly producer services (Bailly and Maillat, 1986). Thus, "beyond the obvious expansion of service activities (especially in terms of numbers of jobs), it is important to realize that the future does not lie in the 'tertiary', if by this is meant the range of activities which is supposed to replace industry as industry once replaced agriculture, but it lies on the contrary in reciprocal synergetic ties developed between the services and the rest of the economy" (Preel, 1986). New service functions must be added to all levels of production networks both upstream and downstream from the manufacturing function (research, management, marketing) (Bailly and Maillat, 1986; Lambooy and Tordoir, 1985; Preel, 1986). This evolution reflects a need to establish ties between the informational and physical aspects of production. "Information is giving structure to physical elements in a production system, combining, co-ordinating, and controlling relations. It plays an integrative role, integrating for example processes in the environment of a firm with its internal (physical) operations. Therefore, professional services form an intermediary or interface between the internal 'hardware' of a firm - the sphere of production - and the external world" (Lambooy and Tordoir, 1985).

The development of such tertiary skills and know-how must not be overlooked by industrial regions lest their production systems remain incomplete and thus stand in the way of regeneration. This naturally poses the problem of the location of producer services. It has often been observed that services and secondary activities develop within different spatial parameters (Malecki, 1983).

Although producer services were for a long time located mainly in large cities, today this is no longer invariably the case. These services are now liable to develop at other levels of the urban hierarchy (Bailly, Maillat and Rey, 1984). This evolution is particularly significant in view of the fact that the areas in question harbour numerous SMEs. Being unable to develop internally all the service functions they need, these enterprises must resort to external aid. There is consequently a growing demand for producer services. This demand is bound to be all the greater as product innovation and vitality are essential to the survival of SMEs. Consequently, the market area within which producer services remain cost-effective is shrinking. This tendency should promote greater proximity between industries and producer services which are capable of constituting useful additions to traditional networks (Lambooy and Tordoir, 1985).

THE LABOUR MARKET

The "mobility channel" concept

The labour market appears to play an essential role in the potential of industrial regions since it represents, through the nature and skills of its labourforce, a regional reserve of human capital, that is of one or several types of specific know-how. Moreover, employment is a determining factor in the regional integration of each individual.

The various existing approaches to regional development all focus on local human capital, that is on the talent, initiative and knowledge of the inhabitants of a region. These human resources are considered capable of providing or triggering the development of regional comparative advantages (Coffey and Polese, 1984). The knowledge and skills embodied in the population have a determining influence on regional innovative capacity. Such human capital resources, consisting particularly of specific skills, promote the development and implementation of innovative solutions (Gaffard, 1986).

An assessment of the relative importance of the spectrum of skills present in a region is often made in an attempt to bring to light differentials in regional capacities (Malecki, 1983). However, this approach falls short of providing a satisfactory understanding of regional variations in innovative capacity. This is because it fails to take into account the mobility of human capital, namely its ability to move from one region to another. In fact, it is a region's capacity to hold on to its human capital which is crucial. This capacity is a function of the number of employment opportunities provided by a region. Thus, the size, structure and characteristics of regional labour markets, in terms of such considerations as required skills, promotion prospects, trainee positions and work conditions, all constitute factors of attraction or repulsion.

The concept proposed here for analysing the capacity of regional markets to retain their labour forces is that of a "mobility channel" (Held and Maillat, 1984; Maillat, 1984; Destefanis, 1983). This concept refers to the manner in which various categories of regional employment evolve and develop, as well as to how they interact and complement each other in such a way as to enable the labourforce to change jobs and transfer know-how from one job to another without having to leave the region.

Regional labour market structure is determined by variations in job content and stability, the role played by jobs in career paths and the characteristics of the employed labourforce. Within this structure, jobs are allocated according to a number of rules, and labour market functioning depends upon the way in which this allocation takes place. The allocation process, defined here as a "mobility channel", gives rise to labourforce flows between various job categories.

Members of the labourforce face a variety of possibilities during their working lives. They may seek promotion within an enterprise (in the case of an internal market), move from one post to another in an effort to increase their professional skills and therefore their chances of finding suitable stable employment, remain in their first-job posts, change jobs frequently without acquiring particular benefits, move often from one precarious temporary job to another, or obtain stable employment on the strength of previous professional experience (second-job posts, with or without promotion prospects) (Held and Maillat, 1984).

Workers thus occupy a variety of jobs in the course of their careers in an order determined by enterprise requirements (recruitment selectivity) and internal promotion rules. Mobility channels may be roughly divided into the three following major categories:

- Vertical mobility channels: workers may enter enterprises and build careers within them, or occupy several jobs in different enterprises in a bid to accumulate experience; that is, in more general terms, to acquire skills of sufficient negotiable market value ultimately to lead to the occupation of very attractive posts in terms of duties, wages and security.
- Lateral mobility channels: the occupation of several posts does not help workers significantly to improve their market value.
- Job entry/exit flows: the absence of opportunities or motivations for changing employment reduces mobility to job entries and exits. The length of job occupancy may vary from several days to several years.

The mobility channel concept shows that employment in itself is less important to a region than the existence of certain job combinations. The presence of such combinations both enables the inhabitants of a region to achieve their ambitions and serves to attract outside interest. The existence and multiplication of vertical mobility channels appears unquestionably as a crucial factor in ensuring an attractive labour market structure. To be more precise, the longer the vertical mobility channels, the greater the region's power of attraction. For unskilled workers, numerous possible moves within lateral mobility channels appear less important than opportunities for stable employment.

The role of establishments

According to the mobility channel concept, jobs cannot be considered independently from the enterprises upon which their existence depends. The specific nature of local labour market structures thus hinges on the types of establishments existing within a region.

Not all establishments offer the same types of employment. Some specialize in the recruitment of young workers to whom they offer training and promotion prospects, others train workers who plan to leave at the end of their training period, others hire only experienced personnel, and still others recruit unskilled workers of all ages, professional backgrounds and degrees of training. Naturally, certain establishments offer several different types of employment.

Ultimately, local labour market structures are determined by the nature of the establishments set up within a region. These establishments constitute stations along the mobility channels.

The direction taken by mobility flows between different types of establishments is not haphazard. They move, for example, from training and first-job establishments towards final-job establishments specializing in the recruitment of experienced or mature personnel. Less type-cast establishments are found along the intervening path. Thus, knowledge of the typology of regional establishments provides a fairly precise idea of how local labour markets function, and helps forecast the consequences of any modifications within these structures.

For example, the disappearance of final-job establishments or even the slightest decrease in the attractiveness of the conditions they offer may suffice to threaten recruitment by other enterprises. This is because disenchantment with meagre regional promotion prospects prompts a certain number of labour-market enterers to seek employment elsewhere from the onset of their working lives. Inversely, a lack of trainee positions or, to an even greater extent, a lack of entry-level posts can radically obstruct the recruitment efforts of other establishments (Maillat, 1984).

Thus, differences exist between regional labour market structures which explain variations in their ability to retain human capital, particularly to hold on to workers who possess specific regional know-how. The existence of regional mobility channels enables workers changing establishments or enterprises to maintain, transfer and increase regional know-how. Mobility is thus possible not only between, but also within regions. The existence of mobility within a region enables it to preserve, to a large extent, its potential know-how since workers, moving from one establishment to another, constitute vectors of that know-how.

Regional mobility chains and establishments: the Neuchâtel example

A useful example of the relationships discussed above is provided by a Neuchâtel region case study in which employment patterns were followed between 1970 and 1979. Some 13 different types of establishments were found to co-exist in the region, as described in table 8.1.

The majority of these establishments suffered from the recession which affected the region after 1970. However, the degree of impact varied considerably from one establishment to another. The most negative impact was experienced by establishments offering the most precarious (type 9) and least skilled (type 11 - mainly female) employment. Establishments with internal markets and those placed along vertical mobility channels fared much better. In general, the ability of an enterprise to withstand the recession was found to be proportional to the extent of skilled employment and promotion prospects it offered. There was undoubtedly a substantial overall job loss during the period in question. However, the various types of establishments constituting the mobility channels of the Neuchâtel region have survived. It can thus be concluded that the recession did not succeed in totally disorganizing mobility channels, and that this allowed the local labour market to preserve a certain degree of its potential know-how as well as to retain part of its labourforce which was endowed with specific skills. This is particularly significant in view of the fact that workers of the latter type can be integrated into new advanced-technology establishments which graft themselves onto existing mobility channels.

Table 8.1 Typology of establishments in Neuchâtel

(1) First-job establishments offering promotion prospects for qualified personnel (internal market up to higher management);

(2) Internal market for qualified personnel, but limited to junior management;

(3) Second-job establishments for managerial and qualified personnel on vertical mobility channels;

(4) Final, stable establishments for qualified personnel;

(5) Internal market for semi-qualified and unqualified men;

(6) Final establishments offering promotion prospects for semi-qualified and unqualified men;

(7) Final, stable establishments for semi-qualified and unqualified men;

(8) Second-job establishments for semi-qualified and unqualified workers with a large stable core and high turnover;

(9) Establishments offering precarious employment for aliens;

(10) Neutral and mixed establishments;

(11) Establishments offering regular, steady employment for semi-qualified and unqualified women;

(12) Establishments offering jobs for women, with a high turnover;

(13) Establishments offering a high proportion of precarious jobs for women, tied to the labour market.

Source: Maillat (1984b)

THE SCIENTIFIC FRAMEWORK

The preceding section showed the important influence exerted by the labour market on regional potential through the development of specific know-how within regional production frameworks. In introducing the concept of a scientific framework, a distinction must be drawn between empirical and analytical know-how (Barcet, Le Bas and Mercier, 1985). Empirical know-how stems from the practical relationship existing between the worker and the object or means of production. Analytical know-how, although invariably bearing a relationship to the production process, implies the additional existence of an ability to carry out a relatively-thorough scientific analysis of the factors involved in that process. This distinction underlies the concept of a scientific framework constituting a source of specific skills necessary for revival. The scientific framework may be defined as the set of relationships existing between science, industry, education and the State (Stohr, 1986). The present analysis focusses mainly on the role of training and research establishments, particularly their contribution to the regional co-production of technology.

Analytical know-how (the incorporation of scientific analysis into the production process) requires a basic foundation of technical knowledge ordinarily acquired through the training system. The connection between a high degree of skills at the enterprise level, which implies a high-calibre training system, and the development of new products and processes seems widely supported by a variety of studies on the subject. A large proportion of highly-skilled scientific and technical personnel within the labour-force structure of enterprises is often a criterion of the presence of vanguard activities (Premus, 1983; Camagni, 1986; Planque, 1986).

The types of skills, frequently quite specific, required by activities conducive to innovation are of vital importance to a regional economy. Although such activities are often presented as footloose, they frequently appear tied to the specific location of the types of labourforce they require. Therefore, a particularly important factor in the ability of a region to cope with technological change is the existence of a territorialized regional training system, complementary with the type of labour market structuring described earlier, and offering a high level of skills.

As for research establishments, whether university-connected, public or private, their role and contribution to the regional production of technology are more controversial. Often credited with the extraordinary development of regions such as Silicon Valley or Route 128, research institutes nevertheless often seem to offer more imagined than real advantages. Meyer-Krahmer (1985) thus shows that in Germany, universities, polytechnic schools and research institutes (public and private) do not play as prominent a role in providing SMEs with know-how as is commonly believed. A study of the Netherlands showed no significant correlation between the location of research institutes and the presence of innovation. Only firms that resorted exclusively to external R & D services tended to obtain external assistance and advice within close proximity (Dieperinck and Nijkamp, 1984). In light of these facts, Dieperinck and Nijkamp concluded that policies aimed at reinforcing regional innovative potential through the creation of additional scientific and technical infrastructures, such as new research centres, would fail to have any considerable impact. Moreover, the development of new means of communication fostering increasingly decentralized access to information may constitute an advantage for regions and enterprises heretofore cut off from information, and may thereby tend to cancel the proximity effect (Planque, 1983).

These observations, although highly relevant in many cases, do not seem to apply to Europe's traditional industrial regions. On the contrary these regions, rocked by technological change and a growing international division of labour, reveal a mounting need for scientific, technical and commercial information concerning the incorporation of innovation into their fields of concern. Although on the one hand some enterprises, generally large ones, succeed in obtaining this information regardless of their location, while on the other hand many enterprises fail out of habit to seek it, neither is true of regional SMEs which are in a position to become a launching pad for the conversion and transformation of regional structures.

Yet these SMEs, although open to technological change, often fail to encompass the entire range of enterprise functions, in particular R & D. The proximity of research centres can play an important part in facilitating the access of SMEs to this kind of information. However, for this proximity effect to be felt, there must be a concomitant focussing of the local scientific network on regional activities as well as a setting-up of interfaces, or research-industry ties. Local scientific networks must be able to secure information, draw in outside technology (through procurement, transfers, exchanges, etc.) and disseminate both of these through regional channels within a coherent industrial system. Several enterprises of a similar nature or belonging to the same mobility channel must exist within close proximity or at least within the same region for there to be a development of the strategic synergetic ties necessary for an active circulation of information. Scientific networks can thus become an integral part of territorial production systems.

CONCLUSION

The above analysis has identified the economic and territorial conditions for the revival of Europe's traditional industrial regions in relation to three elements which appear essential, namely the production network, the labour market and the scientific network. There seems to be a clear need to modify regional policy in consequence.

Above all, regional policy must seek to reconstitute an integral and coherent territorial production system. Its integrity depends upon the presence of service and manufacturing activities organized in a series of upstream and downstream sequences. Its coherence hinges on the existence and synergetic interaction of ties between various territorial system functions (networks, interfaces, etc.). Measures undertaken should aim to revitalize and reinforce the role of SMEs, as active vectors of transformation, in such a way as to recreate opportunities for the evolution and development, through existing or future networks, of the vital factor constituted by regional know-how.

REFERENCES

Amin A. and Thwaites A. T. (1986) Technical Change and the Local Economy: the Case of the Northern Region (UK). In Aydalot P. (ed.), Milieux innovateurs en Europe, GREMI, Paris, pp. 129-61

Andersson A. (1985) Creativity and Regional Development, Papers of the Regional Science Association, 56, pp. 5-20

Aydalot P. (1984) Technologies nouvelles et développement

territorial. Rapport préliminaire et projet de recherche, Paris, Centre Economie-Espace-Environnement

Aydalot P. (ed.) (1986) Milieux innovateurs en Europe, GREMI, Paris

Bailly A. and Maillat D. (1986) Le secteur tertiaire en question, ERESA, Geneva

Bailly A., Maillat D. and Rey M. (1984) Tertiaire moteur et développement régional: le cas des petites et moyennes villes, Revue d'Economie Régionale et Urbaine, 5, pp. 757-76

Barcet A., Le Bas C. and Mercier C. (1985) Savoir-faire et changements techniques, Presses Universitaires de Lyon, Lyon

Boulianne L. and Maillat D. (1983) Technologie, entreprises et régions, Georgi, Saint-Saphorin

Camagni R. and Rabellotti R. (1986) Innovation and Territory: the Milan High-Tech and Innovation Field. In Aydalot P. (ed.), Milieux innovateurs en Europe, GREMI, Paris, pp. 101-25

Castells M. (ed.) (1985) High Technology, Space and Society, Sage Publications, Beverly Hills

Castells M. (1985) High Technology, Economic Restructuring, and the Urban-Regional Process in the United States. In Castells M. (ed.), High Technology, Space, and Society, Sage Publications, Beverly Hills

Coffey W.J. and Polese M. (1984) La localisation des activités de bureau et des services aux entreprises: un cadre d'analyse, Revue d'Economie Régionale et Urbaine, 5, pp. 717-30

Cunat F. (1986) Stratégies d'adaptation d'entreprises du Nord-Pas-de-Calais. Communication à la Table-Ronde CREUSET, St Etienne, May 1986

Daynac M., Millien A. and Cunat F. (983) Politique économique et développement local: modalités et résultats des tentatives de reconversion des zones industrielles en crise. Communication au colloque RSA-ASRDLF, Poitiers, September 1983

Destefanis M. (1983) Le fonctionnement du marché de l'emploi au niveau local en France. In Maillat D. (ed.), Le fonctionnement du marché de l'emploi au niveau local, Georgi, Saint-Saphorin

Dieperink H. and Nijkamp P. (1985) Spatial Dispersion of Industrial Innovation: A Case Study for the Netherlands. Communication au colloque de l'ASRDLF, Marrakech,

1985

Gaffard J. L. (1986) Restructuration de l'espace économique et trajectoires technologiques. In Aydalot P. (ed.), Milieux innovateurs en Europe, GREMI, Paris, pp. 17-27

Hall P. (1985) Technology, Space, and Society in Contemporary Britain. In Castells M. (ed.), High Technology, Space and Society, Sage Publications, Beverly Hills, pp. 41-52

Held D. and Maillat D. (1984) Marché de l'emploi, Presses Polytechniques Romandes, Lausanne

Lambooy J. G. (1984) The Regional Ecology of Technical Change. Paper to the U.N. Conference, Warsaw, June 1984

Lambooy J. G. and Tordoir P. P. (1985) Professional Services and Regional Development: a Conceptual Approach. Paper to the Fast-Conference, Brussels, October 1985

Maillat D. (ed.) (1983) Le fonctionnement du marché de l'emploi au niveau local, Georgi, Saint-Saphorin

Maillat D. (1984a) Les conditions d'une stratégie de développement par le bas: le cas de la région horlogère suisse. Revue d'Economie Régionale et Urbaine, 2, pp. 257-74

Maillat D. (1984b) Les chaînes de mobilité: instrument d'analyse et de gestion du marché de l'emploi au niveau régional, Revue Internationale du Travail, 123, 3, pp. 379-94

Maillat D., Schoepfer A. and Voillat F. (eds) (1984) La nouvelle politique régionale; le cas de l'arc jurassien, EDES, Neuchâtel

Maillat D. and Vasserot J.-Y. (1986) Les milieux innovateurs; le cas de l'arc jurassien suisse. In Aydalot P. (ed.), Milieux innovateurs en Europe, GREMI, Paris, pp. 217-46

Malecki, E.J. (1983) Technology and Regional Economic Change: Trends and Prospects for Developed Economies. Paper to RSA-Meetings, Chicago, November 1983

Martin F. (1986) L'entrepreneurship et le développement local: une évaluation, Revue Canadienne des Sciences Régionales, printemps, pp. 1-24

Meyer-Krahmer F. (1985) Innovation Behaviour and Regional Indigenous Potential, Regional Studies, 19, 6, pp. 523-34

Nishioka H. (1983) High Technology Industry Location and Regional Development. Paper to OECD-Workshop, Paris, October 1983

Oakey R.P., Thwaites A.T. and Nash P.A. (1980) The Regional Distribution of Innovative Manufacturing Establishments in Britain, Regional Studies, 14, 3, pp. 235-53

OCDE (1984) Rapport analytique sur la recherche, la technologie et la politique régionale. OCDE, Paris

Perrin J. C. (1983) La reconversion du bassin d'Alès; contribution à une théorie de la dynamique et de la politique locale, Notes de recherches du CER, No. 12 - 1983/2

Perrin J. C. (1984) Dynamique locale, division internationale du travail et troisième révolution industrielle, Notes de recherches du CER, No. 50 - 1984/10

Peyrache V. (1986) Mutations régionales vers les technologies nouvelles; le cas de la région de Saint-Etienne. In Aydalot P. (ed.), Milieux innovateurs en Europe, GREMI, Paris, pp. 195-215

Planque B. (1983) Innovation et développement régionale, Economica, Paris

Planque B. (1986) Pôles d'innovation, PME et planification régionale. Rapport final, Aix-en-Provence, CER

Pottier C. (1984) The Adaptation of Regional Industrial Structures to Technical Changes. Paper to RSA-Congress, Milan, August 1984

Preel B. (1986) Pour "servir" les entreprises que peuvent faire les collectivités locales? Espaces Prospectifs, 4, pp. 83-116

Premus R. (1983) Location of High Technology Firms and Regional Development. Paper to OECD-Workshop, Paris, October 1983

Rothwell R. and Zegveld W. (1982) Innovation and the Small and Medium Sized Firm, Frances Pinter, London

Stohr W.B. (1986) Territorial Innovation Complexes. In Aydalot P. (Ed.), (1986), Milieux innovateurs en Europe, GREMI, Paris, pp. 29-54

Thomas J.N. (1986), Innovation et territoire. Communication à la Table-Ronde CREUSET, St Etienne, May 1986

Vasserot J.-Y. (1986) L'encouragement économique des régions en Suisse: l'exemple de l'arc jurassien. IRER, Neuchâtel

Chapter 9

Local Employment, Training Struc New Technologies in Traditional Indus European Comparisons

Mateo Alaluf and Adinda Vanheers

A survey recently repeated after an interv
Lorraine steel company clearly highligh
influence of technological innovation in d
areas in Europe. Whereas 20 years ago
agreed that technological progress was
workers' only concern was how to shar
today, with economic crisis and job
phenomenon is viewed as a source of
poverty (Durand, 1980). Yet at the
technology is seen as providing a possib
employment crisis caused by the e
industries.

Five sunset European industrial regio
the work of GREMI, as outlined in chap
giving a clearer picture of the soc
technological innovation in declining in
amassed through the systematic obser
canton (Switzerland), Besançon and St
Charleroi (Belgium) and the Northern Re
(Aydalot, 1986, 129-266). An attempt ca
pick out evolutionary patterns w
distinguishable from spellbound visions
miracles" generated by crisis situations.

STRUCTURE OF ACTIVITIES

The five areas can be divided into tw
first, three poles of heavy indus
(Charleroi, the Northern Region and
characterised by a marked concentrati
capital, around which a group of smalle
involved in sub-contracting work for

Oakey R.P., Thwaites A.T. and Nash P.A. (1980) The Regional Distribution of Innovative Manufacturing Establishments in Britain, Regional Studies, 14, 3, pp. 235-53

OCDE (1984) Rapport analytique sur la recherche, la technologie et la politique régionale. OCDE, Paris

Perrin J. C. (1983) La reconversion du bassin d'Alès; contribution à une théorie de la dynamique et de la politique locale, Notes de recherches du CER, No. 12 - 1983/2

Perrin J. C. (1984) Dynamique locale, division internationale du travail et troisième révolution industrielle, Notes de recherches du CER, No. 50 - 1984/10

Peyrache V. (1986) Mutations régionales vers les technologies nouvelles; le cas de la région de Saint-Etienne. In Aydalot P. (ed.), Milieux innovateurs en Europe, GREMI, Paris, pp. 195-215

Planque B. (1983) Innovation et développement régionale, Economica, Paris

Planque B. (1986) Pôles d'innovation, PME et planification régionale. Rapport final, Aix-en-Provence, CER

Pottier C. (1984) The Adaptation of Regional Industrial Structures to Technical Changes. Paper to RSA-Congress, Milan, August 1984

Preel B. (1986) Pour "servir" les entreprises que peuvent faire les collectivités locales? Espaces Prospectifs, 4, pp. 83-116

Premus R. (1983) Location of High Technology Firms and Regional Development. Paper to OECD-Workshop, Paris, October 1983

Rothwell R. and Zegveld W. (1982) Innovation and the Small and Medium Sized Firm, Frances Pinter, London

Stohr W.B. (1986) Territorial Innovation Complexes. In Aydalot P. (Ed.), (1986), Milieux innovateurs en Europe, GREMI, Paris, pp. 29-54

Thomas J.N. (1986), Innovation et territoire. Communication à la Table-Ronde CREUSET, St Etienne, May 1986

Vasserot J.-Y. (1986) L'encouragement économique des régions en Suisse: l'exemple de l'arc jurassien. IRER, Neuchâtel

Chapter 9

Local Employment, Training Structures and New Technologies in Traditional Industrial Regions: European Comparisons

Mateo Alaluf and Adinda Vanheerswynghels

A survey recently repeated after an interval of 20 years in a Lorraine steel company clearly highlights the compelling influence of technological innovation in declining industrial areas in Europe. Whereas 20 years ago it was generally agreed that technological progress was beneficial and the workers' only concern was how to share out the profits, today, with economic crisis and job loss, the same phenomenon is viewed as a source of unemployment and poverty (Durand, 1980). Yet at the same time, high technology is seen as providing a possible solution to the employment crisis caused by the ebb of traditional industries.

Five sunset European industrial regions were included in the work of GREMI, as outlined in chapter 1. Information giving a clearer picture of the social conditions of technological innovation in declining industrial areas was amassed through the systematic observation of the Jura canton (Switzerland), Besançon and St Etienne (France), Charleroi (Belgium) and the Northern Region (Great Britain) (Aydalot, 1986, 129-266). An attempt can thus be made to pick out evolutionary patterns which are clearly distinguishable from spellbound visions of "mirages and miracles" generated by crisis situations.

STRUCTURE OF ACTIVITIES

The five areas can be divided into two sub-groups. In the first, three poles of heavy industrial development (Charleroi, the Northern Region and St Etienne) were characterised by a marked concentration of industry and capital, around which a group of smaller companies largely involved in sub-contracting work for the larger firms gravitated. The second consists of two centres of traditional light industry, namely clock and watch-making, the origins of which date back to the 17th and 18th centuries. The industrialisation period of these two was marked for the most part by the emergence of small and medium-sized firms built upon family capital.

Regions of heavy industry

The industrial development of the first three regions, which was hinged upon well-ingrained traditional craft industries (metal-work, nail-making, cutlery-making, arms manufacture, glass-work, shipbuilding, ribbon-making, textiles) was notably linked to the presence of raw materials, among which coal played an essential role. These three areas form part of the group of European regions which were industrialised very early on and in which a number of major technical innovations were successively introduced to increase coal extraction capacities, to improve the performance of naval and rail transport, and to develop sectors such as textiles, steel and glass-making. These regions were centres of attraction for and concentration of local and external industrial capitalism. Up until the 20th century, they were characterised by a degree of integration and intense interaction between local industries. The main motors of their industrial activity were broadly as follows:

- Charleroi: glasswork, steel, electrical engineering;
- Northern Region: steel, shipbuilding;
- St Etienne: steel, textiles.

The sectors of economic activity which were the driving forces behind the wealth of these centres have also been responsible for their decline. Despite the slump in employment levels, a large proportion of the industrial work-force is still concentrated in these regions. The dismantling of formerly strong sectors has gone hand in hand with a trend towards the concentration and the internationalisation of industrial capital. As a result, regional economies have fallen under the domination of large multi-plant corporations. Activities have either been redirected or restricted and local companies have closed down or faded away. The (de)localisation policies arising from the economic development strategies of multinationals, over which local people have no influence, are responsible for this. The regional context is therefore no longer a synonym of integration but becomes in some ways an important implantation site for large firms and for small and medium-sized enterprises (SMEs). The latter compete against one another and against external forces whether they are involved in sub-contracting work or operating in a totally autonomous fashion. In such a situation, the local environment no longer embodies of its own accord the factors essential for recovery and economic dynamism.

Areas of light industry

On the other hand, in the Swiss Jura canton and Besançon, small and medium-sized concerns grew up, concentrated in clock and watch-making and related local activities. Besançon, as capital of the region, also has a highly developed tertiary sector, which is not the case in the Jura canton. The clock-making industry began to experience growing problems in the Besançon region in the middle of the 1970s, and the other manufacturing industries followed it into decline. Reconversion then took place in the watch-making industry. Large industrial corporations intervened, external capital gained an upper hand and the area adopted micro-electronic technological innovations. The metalwork sector reconverted to the production of electronic components and subsidiaries of external groups came to set up business in the region, attracted by the high quality of local sub-contracting work.

In the Swiss region, the manufacturing sector was predominant and essentially revolved around clock-making and the machine tool industry. It also began to decline at the beginning of the 1970s. Again, fresh economic stimulation in the region was provided by foreign companies, along with steps taken to encourage the development of local companies. These two areas, which at the outset had very similar mono-industrial structures, have evolved towards more varied and heterogeneous activities, developed partly by foreign companies or by the subsidiaries of externally-owned firms. However, this evolution has led to a slackening of inter-company relationships due to more intense external connections and direct competition from companies on the international market.

EMPLOYMENT SITUATION

Regions of heavy industry

The three areas studied all experienced a significant drop in employment in traditional sectors which was not compensated for by growth in the tertiary sector or by job creation in "new" activities (either because implantation is recent, or because it has not taken place in the traditional heavy industry regions). Unlike the Northern Region, the tertiary sectors of Charleroi and St Etienne are less developed than the national average.

In the Northern Region, industrial employment is indeed concentrated in a small number of companies involved in the manufacture of metal goods and machine tools, followed by the food, chemical, oil, textiles and coal industries. External firms have majority holdings in these concerns and when they are active in crisis sectors, they operate a policy of trimming the workforce, as is seen in the fact that the major part of job losses in the manufacturing sectors can be attributed to them. The introduction of technological innovations in these local subsidiaries results in rationalisation of production and release of staff. In addition, smaller companies do not make a very significant contribution to the local economy. Although their relative employment proportion is growing, in perspective this is only because of loss of employment in major firms.

The level of unemployment in these regions is currently very high, coming close to 19%. This figure is the net result of a slump in male employment in manufacturing industries and some growth in female employment in services.

The management policies of the major corporations cause delocalisation of research activities or management functions to other areas, the shut-down of profitable companies after their take-over and lower pay for staff.

In the Charleroi region, the highest concentration of workers is still to be found in the traditional sectors of activity, despite substantial workforce reductions. The level of unemployment in the region is high and includes a considerable proportion of long-term unemployed.[1] The situation in the region has deteriorated rapidly since the 1950s due to the crisis in the coal industry followed by decline in the engineering, glass-making, electrical engineering and steel sectors. However, awareness of regional decline has been partially masked by the daily "shuttling" of workers to other regions in the country and by the drop in the activity rates of the population in this region, particularly in female activity rates.

The last time a big company set up business in the region was almost 15 years ago and the only recent creations have been SMEs. Although in this region a great deal of effort goes into encouraging the creation or development of local companies, this only leads to a very small employment increase. In addition, it often concentrates on the hiving off of companies or on sub-contracting networks revolving around the major corporations operating in the region. It is more a question of increasing competition between local sub-contractors than of setting up inter-industrial networks. Moreover, hiving off and recourse to sub-contracting companies often involve substitution for certain activities which were previously carried out within major companies. Once again, this is the result of the management and development policies of large firms.

In the St Etienne region many companies have been set up in the 1980s. Those which register the fastest employment increases are centred, in the tertiary sector, on activities such as "studies, advice and technical assistance", "social work", and "wholesale trade"; and in the secondary sector, on "the agro-food industry", "metalwork", "machine woodwork" and "the production of industrial equipment". These encouraging results are however at present only compensating in a very partial manner for job losses in traditional sectors.

Regions of light industry

Besançon and the Swiss Jura canton have a certain number of common employment characteristics which can be attributed to activity similarity and to a comparable traditionally capitalistic structure. They are however distinguished by a different sectorial activity spread. Although there is a strong tertiary sector in Besançon, the same cannot be said for the Jura canton.

In the last ten years, the Jura canton has lost a considerable number of jobs in its main industrial branches, namely clock-making and the machine tool industry. This situation has been followed by negative demographic growth both of the resident and the working population.

The Besançon region, on the other hand, has a working

population which is younger than the French average and a high degree of female employment concentrated primarily in the tertiary sector. It only began to feel the impact of declining manufacturing employment at the end of the 1970s, since the growth in the tertiary sector had a considerable buffer effect. However, since the mid-1970s, more than 50 per cent of jobs in the watch-making sector have been lost. By 1980, employment in the manufacturing sector generally was also declining rapidly. Regional unemployment levels, which were relatively low in the 1970s, have consequently grown more rapidly than the national average since 1980.

In regions with sunset industries the question of the impact of technological innovations upon employment is especially significant. The industrial structures which, since the beginning of the eighteenth century, have benefitted from technological progress made these regions the focal point of an exceptional development phase. The evolution of the capitalist process which accompanied this industrialisation and in particular the concentration and internationalisation which contributed to the wealth of these industrial areas and increased the complexity of local economic structures are also at the heart of the profound structural crisis which they are currently experiencing.

TECHNOLOGY AND EMPLOYMENT

Although their form and scale has not always been the same, the different phases of restructuring which have successively taken place since the beginning of the 1970s in these regions have caused major job losses. In addition, innovations in manufacturing processes bring about labour savings. Product innovations or the development of new high technology activities only provide a very meagre compensation for these job losses. Despite its importance, the direct quantitative contribution to employment of the incorporation of new technology in production processes is not comparable with the cuts in employment which it provokes.

Whatever the nature, consequences and range of the modifications involved in different European regions, it has been demonstrated on a number of occasions that existing local structures generally adapt much better to the changes than is usually imagined. This is as true for regions of heavy industry as it is for regions of light industry.[2]

How, then, can the changes observed in the regions studied be characterised? Some of the developments seem to be a consequence of the characteristics of the companies and their organisation modes which directly affect employment structures.

Firstly, company sizes and organizational characteristics are changing. Small and medium-sized units are being set up and sub-contracting is undergoing growth. "Alternative" companies are also being created, introducing self-management or new co-operatives. These developments do not however seem to be hindering the concentration trend which has hitherto been evident in regions with sunset industries. They are on the contrary a demonstration of the capacity of large corporations to adapt to the changing industrial environment, which is characterised by more rapid product and process innovation, and shorter product and process life expectancy; in other words, by continually regenerative transformation. The company diversification and dependency networks described notably with reference to Charleroi or the Northern Region (Aydalot, 1986, pp. 163-94) thus also reflect, in our opinion, the modern use made of SMEs by large firms.

These changes concern not only the company profile (size, age, life-span, mobility) but also its internal organisation system, which has also undergone transformation. Change has been seen in a number of areas, including increased decentralisation of management, modification of career plans, adoption of "matrix" organisational systems resulting in two-dimensional executive-hierarchy relationships while preserving the coherence of the central management and services,[3] and use of new forms of work organisation such as autonomous or semi-autonomous teams, "quality circles" and so on.

As regards the structure of employment in the declining industrial regions under discussion here, even though tertiary activities are under-represented, in general the proportion of managerial and higher-level staff is increasing steadily relative to that of production workers as a proportion of total people employed. This evolution is a sign of an even more profound shake-up of employment structures, arising from new patterns of work organisation and placing the emphasis on intellectual work.

The normal explanations given for new work organisation forms are the desire for more flexibility, higher product quality, stock reduction and activation of worker capacity. However, there is no linear evolution in production automation. Successive generations of machines and equipment co-exist in the same companies. The same applies to heterogeneous forms of work organisation. Even if the idea of complete company integration seems to find its contradiction in its own complexity, the arrangement and control of computerised data flows in fact constitutes, behind the apparent organisational patchwork, the central nerve of new producer companies. This process, which optimises more "intellectual" skills, is evident not only within big companies but also affects the body of relationships between designers and assemblers, sub-contractors and distributors. In other words, it concerns the entire local economic unit.

The worker-work relationship which became dissociated with the evolution from craft to industry is taken one step further apart by automation. The time of operation execution by human operators is disconnected even further from machine time. This severing of worker activity from machine operation is also becoming increasingly widespread at the local level due to the division of work between companies and branches of activity. As a result, as Pierre Veltz has pointed out, there is a "growing dissociation in time and in space between the physical transformation of matter and human action" (Veltz, 1986: 18). A worker no longer just transforms matter, but also and increasingly has to handle information, design and organisational activities. A single workshop is no longer enough. Work activities are a function of the multiplication of information flows around the factory, which bring to the fore inter-company relationships, capacities and regional research and investment support networks.

SCHOOL AND WORK

Information has therefore become a strategic industrial raw material. This phenomenon has led not only to an increase in the number of engineers and technicians, but also to the transformation of work in its entirety. A growing number of workers, both on the shop floor and in offices, now use computer terminals. But what does this shake-up represent in terms of qualifications and training systems?

The separation of human and mechanical (or automatic) operations prevents, in our opinion, any simple transposition of the technological transformations into work conditions and even less into training systems. How have training systems evolved in relation to economic structures?

In regions of heavy industry which grew up around coal, metallurgy, engineering or textiles, there exists a kind of local "know-how" and technical background of the industry in question. Survey replies given by company heads all reveal the existence of this "know-how" in engineering and metal work.

At the dawn of industrialisation, apprenticeship and training for the vast majority of workers took place on the factory floor or in the workshop. Awareness then gradually developed of more technical aspects and of production management. Universities, large schools and industrial and professional colleges assigned themselves the task of training the personnel required by the economy. For example, when, in the 19th century in Belgium the first industrial colleges were set up, they were clustered around the industrial centres.[4] This training was intended to allow workers to become "educated worker, foreman, shop overseer and even, in exceptional cases, good factory manager".[5]

The network of highly developed technical and professional teaching which currently characterises regions such as St Etienne, the Northern Region or Charleroi is rooted in a tradition which dates back to the birth of industrialisation. Technical and professional schooling are therefore also in part responsible for the "know-how" forged by industrial tradition.

In the region of Besançon and the Swiss Jura, the skill of the clock-making workforce became renowned very early on. This know-how was handed down from generation to generation without any significant intervention on the part of the school system. Moreover, the proportion of trained engineers, technicians and supervisors within the production staff is relatively low in these regions.

Of course, it would be an oversimplification to argue that regions of heavy industry introduced worker training in schools whereas regions of light industry handed down technical expertise from family to family. Although the importance and the precocious character of the technical and professional schools in Charleroi has been emphasised, the region's glass industry belonged to the old industrial tradition, yet no preparatory schools exist for this trade. It is handed down from father to son and the family framework remains the training centre for qualifications which cannot be obtained at school.

Certain sectors are now faced with the problem, or have been for quite some time, of transplanting to the school in-company training handed down from generation to generation in working-class families. This form of dependency of private life on working life and the strict relationship between work and individual is coming to an end. In other words, the dissociation of family cell and work cell which results from in-school training is not so much the regrettable ending of a supposedly fulfilling craft profession as a new opportunity for wider training beyond this face-to-face relationship, within the company. It consequently corresponds to an "opening up of social relationships" (Veltz, 1983: 30-42).

SHIFT TOWARDS LOCAL CONSENSUS

Although the various people questioned in the surveys recognised the existence of local "know-how" arising from the specific industrial traditions of the regions, only a few of these thought that this could be a potentially innovative factor, whereas others regarded it as a barrier to modernisation. Know-how can thus involve the perpetuation of jealously guarded rigid techniques which prevent adaptation of procedures and products to market evolution. Equally, know-how is also knowledge, skills and habits upon which innovative projects can rest and which make innovation feasible.

In all the traditional industrial regions mentioned here, the desire to strive towards a local consensus seems to us to be constantly present. In all cases, "know-how" is considered as part of regional identity and the latter is the foundation of the desire for a consensus.

This assertion is substantiated by the action of institutions which attempt to make the most of this "know-how", even if only as a regional trademark, by the activities of trade unions and employers' associations which become involved in projects to reconvert regional sectors, by local employment initiatives, and by training initiatives which aim to make the most of this "know-how". Local identity is the basis of the desire to mobilise the "driving forces" of the region for its development.

In all these regions, a clear trend is now emerging of a search for a socio-economic climate favourable to investment. Recent developments in labour relations would tend to favour the development of such a consensus. With the growth in unemployment, centralised interprofessional or sectorial collective bargaining is giving way to more localized action. Companies and regions are becoming increasingly introspective. In many cases, as in Charleroi, local collective work agreements are being concluded and conferences of "dynamic regional forces" organised with a view to drawing up reconversion projects.

The activity of regional institutions and development agencies, sometimes media-directed, is continually expanding and intensifying. Whereas in the 1960s the main aim was to attract external investors, an effort is now being made to encourage the development of local companies.

What is the significance of this search for a consensus? Have the various factions, senses of identity and of belonging to a corporation been neutralised by recession and economic crisis? Or is this an exercise with a zero net result which will only have the effect of setting regions which have become competitors against one another, as has become the case in the steel industry? Such a consensus could therefore deprive a region of its dynamism, which is the source of its former wealth. Or, on the other hand, are the mushrooming of institutions, the search for forms of co-operation between workers' and employers' organisations, the training experiments and employment initiatives, rather the translation of a desire to open up the region, to diversify relationships between social partners, representing attempts by communities to direct their own future? In real life, all these various aspects seem to be blended together, and the local environment seems at the same time both backward-looking and innovative.

CONCLUSIONS

A number of major observations can be made following this crisscross examination of monographs of traditional industrial regions. They are as follows:

1. Job creation in new technology sectors appears remarkably rapid in most of the regions investigated, but it far from compensates for the job loss in traditional sectors. However it is bringing with it changes in the structure of companies, organisation modes and work forms.
2. The link between these transformations and the content

of the tasks carried out in the production unit is however only hypothetical. The differing practices of companies do not permit any single or simple conclusion on this link, and the question should perhaps remain open by definition. For similar jobs using automated equipment, some companies prefer to employ computer scientists, whereas others take on traditional mechanics on the grounds that traditional techniques remain the operating base of the system. In any event, the gap between operations carried out by man on the one hand and by machine on the other undoubtedly widens very considerably the range of work organisation options for a given piece of equipment.

3. In such a situation, workers' "know-how", or their "technical background" derived from the region's industrial tradition, makes a major contribution to forging a professional identity which merges with local or regional identity. This plays an important role in the attempts to establish a consensus.
4. The increase in the educational level of the working population corresponds to a decrease in the role of the family unit in handing down professional "know-how". This evolution also gives young people the chance to step out of their father's footsteps. The introduction of professional training in education therefore opens up social relationships.
5. Moreover, in all the regions examined, general, technical and professional training were a priority and the source of interesting experiments. A joint training project often forms favourable common ground between social partners and those involved in the social dialogue and thus represents an important element in the quest for a consensus.
6. There is no purely technological innovation in the field of economics. Innovation occurs when coherence is established between technology and the society. The search for this coherence will always be gradual and tentative. Even if the old industrialised regions often want to change their image, marked by the past, they are deeply dependent on it and draw a large part of their dynamism from it.

NOTES

1. In Belgium unemployed persons who satisfy the necessary conditions receive unemployment benefit regardless of the time out of work. The statistics therefore cover all unemployed persons receiving benefit, whatever their period of inactivity.
2. The example of St Etienne can be quoted here. Véronique Peyrache notes that "the traditional sectors are also sectors which can come up with innovative and patentable ideas" (p.200); and Denis Maillat and Jean-Yves Vasserot write with reference to the Swiss Jura canton that "while it is true that the mono-industrial structure based upon watch-making has been the source of major problems, watch-making nevertheless at least indirectly paved the way for new technologies in the region" (p.243); in Aydalot (1986).
3, See on this point the description of the "daisy" organisation at the Caterpillar factory in Charleroi in Aydalot (1986, pp.182-8).
4. Out of 14 listed industrial schools in Belgium in 1866, five were to be found in the province of Hainaut, in which Charleroi is situated. Report on industrial teaching presented to the legislative chamber on April 2, 1867.
5. Source as in note 4.

REFERENCES

Aydalot, P. (ed.) (1986) Milieux innovateurs en Europe, GREMI, Paris

Durand, C. (1980) Les ouvriers et le progrès technique: Mont-Saint-Martin vingt ans après, Sociologie du Travail No.1, pp. 4-21

Veltz, P. (1983) "Fordisme, rapport salarial et complexité des pratiques sociales: une perspective critique, Critiques de l'Economie Politique, Nos 23-4, pp. 30-42

Veltz, P. (1986) Informatisation des industries manufacturières et intellectualisation de la production, Sociologie du Travail, No.1, p. 18

Chapter 10

Development Theory, Technological Change and Europe's Frontier Regions

Remigio Ratti

INTRODUCTION

Europe is - apart from the great Russian plain - a fragmented continent, a space naturally divided into compartments reflecting a political physiognomy which is extremely fragmented in its western and central parts. In spite of these divisions, Europe is very densely inhabited (four times the world's average, excluding the soviet region) and very open to outer and inner exchanges. Frontier regions and their problems are thus very common occurrences in Europe, although they have not often been given attention by the literature on regional science (Hansen, 1983).

Interest and concern with the development problems of frontier regions in Europe have increased considerably since the enforcement of the European agreement on trans-frontier co-operation between territorial authorities and collectivities (Council of Europe, Madrid, 21.5.1980). Even if the political aim pursued by the strategy of the construction of a Europe of Regions is very clearly stated, it is, unfortunately, quite difficult to understand the means whereby these zones may be developed.

On one hand, regional development theory - mainly location theory - suggests that these regions are generally dominated by conflictual relations and are thus hindered in their development process. On the other hand, empirical observation shows that some of them are already very developed (Gaudard, 1971; Ratti, 1971; Hansen, 1983), and even exhibit the characteristics of emergent peripheral regions (Ratti and Di Stefano, 1986). It is therefore necessary to go beyond the simple static vision and examine the development processes related to frontier zones.

Are these development processes merely those common to all types of peripheral region, or do we need a specific theoretical approach for frontier regions? What happens, in particular, to those frontier regions which are at the forefront of current dynamics of technological innovation? Can we penetrate behind locally distinctive features and differences and develop a typology of situations?

Faced with processes of technological innovation, the local environment of frontier regions seems to offer new possibilities in the Europe of the 1980s. In the past, industrial development in such regions tended to be of the "tariff factories" type. Now, however, it is possible to hypothesise a change of nature in the economic meaning of the frontier in the European context. The frontier space, an example of a discontinuous zone depending on the juxtaposition of two or more state economic systems (Ratti, 1981), may be evolving as a better structured area, an integrated periphery. It will thus no longer be characterized by the specificity of its production, but by its functions within a more global process of regional restructuring (Aydalot, 1986).

After a critical analysis of traditional hypotheses concerning frontier region development, this chapter will develop a new hypothesis about the study of these regions in the context of the international division of labour, before concluding with an analysis of the implications of the current phase of industrial and technological restructuring for frontier region development.

THEORIES AND HYPOTHESES OF FRONTIER REGION DEVELOPMENT

Frontier region definition and structural effects

Following Hansen (1977), we choose first a narrow and operational definition of the concept of frontier region. This relates to that part of a natural territory in which economic and social life is directly and significantly influenced by the proximity of an international frontier. We thus consider here only open or potentially open regions, excluding regions with a closed natural border situation, like, for example, some parts of the Alps. However, two or more frontier zones together can be considered as transfrontier regions.

Secondly, it is worthwhile considering the institutional and political situation together with economic factors, following the model of the interrelations between national subsystems set out in Figure 10.1. In this sense, frontier regions have enough special characteristics to justify separate attention, within the wider context of peripheral zones as a whole.

Economic and political elements, in their spatial and historical dimensions, explain the frontier phenomena much better than mere geographic location. A good example here is the case of the movement of foreign workers across the border into Switzerland in the 1960s and 1970s. This can only be understood in the context of such political and economic considerations as the changing nature of Swiss legislation on foreign labour (control and limitation of resident foreigners after 1963; relative liberalization of policy towards border workers after 1966), and theories of polarized growth (the case of Basel and Geneva) and segmentation of the labour market (the case of Jura and Ticino), as discussed in Jeanneret (1985) and Ratti et al. (1982).

Traditional location theory and frontier region penalization

The framework of the relations between politico-institutional and socio-economic subsystems set out in Figure 10.1 and emphasized above helps us to appreciate the contributions of two classic authors of spatial analysis.

Christaller (1933) already accepts the principle of the socio-political separation of the frontier as a third system of spatial organization, along with those based on economic principles of market and transport organisation. For Christaller, however, the principle of political separation can only be considered as an artificial element of imbalance in the common and central market areas. The frontier is thus a factor splitting the hinterlands of central markets because of increased investment costs reflecting higher risks in international border zones. The accumulation of these effects restricts frontier regions to the role of low-order central market places with only a low degree of complementarity and a limited capacity for economic development.

August Lösch (1940) underlines the conflict between political and economic objectives in this type of region: economic priorities and aims - productivity, "Kultur", power, continuity - confront an exactly opposite set of

Figure 10.1 **Frontier regions and the interrelations between national political and economic subsystems**

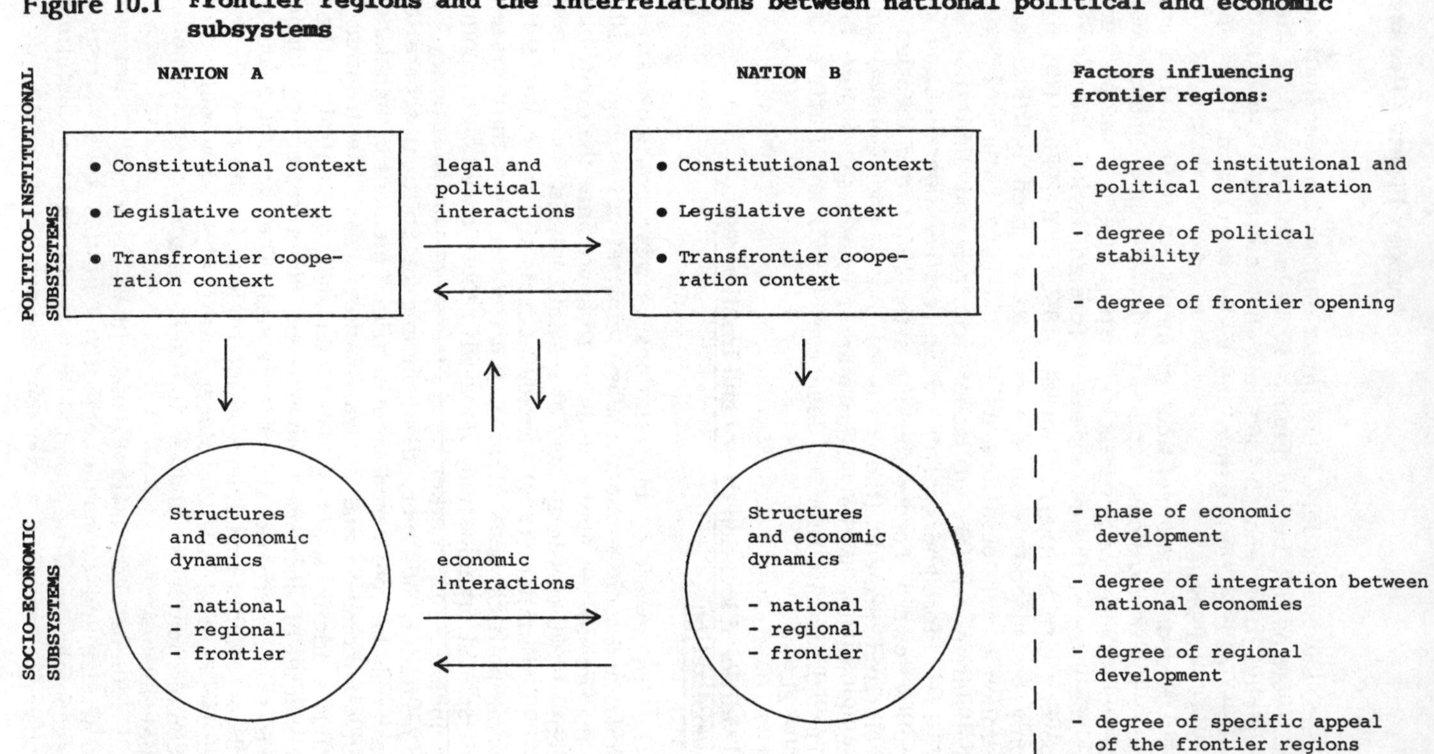

Ratti (1987).

political priorities and aims. The consequences of this, in the form of policies, habits and customs rates, separate complementary economic areas, while public contracts and military objectives introduce real barriers. They induce an additional negative discriminant effect for development of this type of peripheral region.

Admittedly, even in the most traditional theory of location, some favourable impacts of frontier position are acknowledged. Thus an interruption of goods transportation at the frontier - as in the Swiss cases of Basel and Chiasso (Ratti, 1971; Spehl, 1982) - can stimulate particular activities, of great importance for some frontier settlements. Again, in the face of protectionist policies, there has been the implementation of practices like "tariff factories", which are industrial investments in frontier zones intended to penetrate the market of the other nation. Examples here are Swiss investments in neighbouring areas of Germany, Italy and France, and US investments in Canada and Mexico (Peach, 1986). In Europe at least, the latter have now declined following tariff liberalization, while many of the former have been modified by rationalization, economic integration and new technologies (Ratti, 1984; Ratti and Cavadini, 1986).

These remarks highlight the weaknesses of a location theory which is essentially static. Thus while the hypothesis of a penalization effect is plausible, it only appears to be true in those cases where other very important political and economical factors have exerted a negative impact on particular peripheral frontier zones. One example here is the case of Alsace, which suffered economically for many decades as a result of political instability and centralization (Urban, 1971). Another is the Italian-speaking Swiss canton of Ticino, situated to the south of the Alps. This canton was strongly penalized over a long period by the creation of the Swiss Customs Union (which developed from the confederation of cantons to the Federal State in 1848), and later, by the building of the Italian national state after 1861. The negative impact of political centralization was reinforced by that of industrial and urban concentration. In these conditions, even the building of the important European railway through the St Gothard in 1882 only contributed to a very slow change in the economic situation of the Ticino peripheral frontier region (Biucchi, 1959).

To conclude, it may be said that the recent literature on this issue allows us to draw at least a partial conclusion:

the simple finding of the negative effects of a frontier, particularly in the location of production activities, must not be taken as representing a "development theory of frontier regions". Development must be understood, on the contrary, as a joint action - variable in time and space - of both the political and the economic order. It is thus necessary to turn to more suitable and dynamic instruments of spatial analysis.

Frontier region development and the international division of labour

In the broad context of economic growth and increasing liberalization of trade since the World War II, a more attractive theory for the interpretation of the development process of frontier regions arises from more dynamic approaches to the study of location and from the changes undergone by value scales in the regional problem. It is a matter of considering these regions not only in their national space context but also in the larger one of the emergence of a world economy (Michalet, 1976). It is also a question of taking into consideration the concept of the spatial division of labour.

Under present conditions of production and market demand, large firms "do not need anymore to concentrate their activities at only one point, in a huge factory bringing together all kinds of activities and all types of labour. From now on, they are going to split their activities into units which are as homogeneous as possible as far as the labour they employ is concerned, and they will locate these units in places where the necessary labour is available" (Aydalot, 1986).

The result of this process is a territorial organization characterized by a hierarchic and spatial distribution of segments of production, polyvalent by function and by sector. It is thus a more differentiated and complex process than the spatial distribution implied or determined by the phases of the life-cycle of a product (Vernon, 1966).

What then is the role of frontier regions in this process of spatial scattering of activities? Can we identify any special characteristics compared to other peripheral regions? If so, how may we evaluate the consequences for their socio-economic development?

In the context of the spatial and international division of labour model, it is noteworthy that frontier regions - to the extent that they are really open (see our definition) - do appear to exhibit interesting characteristics regarding the possible location of specific and functional segments of activity and production. This is so for three reasons:

- an economic reason linked to a closeness effect; a frontier region is a piece of the national territory and a zone of separation and contact at the same time. It is thus a space of tension, but one which, in a way, anticipates contact with the other country. A possible location in this area - in the national territory or in the border area - can be very attractive because it allows firms to combine proximity advantages (better knowledge, the presence in the region of economic agents with special knowledge of both or many politico-institutional systems) with those determined by the logic of spatial delocation of activities;
- a social reason linked to the flexibility of the regional labour market; the frontier, by its legal and control functions (Guichonet and Raffestin, 1974), creates in an easier way than elsewhere conditions for the control and segmentation of the labourforce (Doeringer and Priore, 1971) in accordance with the requirements of the delocated production units (discrimination by legal or "de facto" measures because of different motivations of labour coming from the other side of the frontier);
- a cultural reason linked to the degree of permeability of the local society; trying to avoid value judgements (positive or negative), one can nonetheless view frontier zones as areas which are more readily permeable for all kinds of circumstances. Factors involved here include greater necessity of adaptation and of taking higher risks, greater frequency of migration and different weights of tradition and identity. The analysis of actor behaviour (Spehl, 1982) and of the perception of the transfrontier reality (Leimgruber, 1984) play an important role in the appearance of frontier effects.

The three reasons listed above remind us immediately of the question of the role, positive or negative, which they play in regional development. For the moment, however, we wish to assess how far these factors may be influencing frontier region development in the context of the new logic of corporate spatial organisation noted earlier. Specifically, many case studies confirm a new thesis about the potential appeal exerted by these frontier regions on the location of activities distributed within the spatial hierarchy of production.

First of all, it is possible in several cases to demonstrate the existence of a significant proximity effect - socio-economic and cultural - that determines favoured relations for the regions situated in each side of a frontier, and which appears to result in above-average levels of regional investment and trade (Jeanneret, 1985). The study by Jeanneret of Swiss zones bordering Austria, the Federal Republic of Germany, France and Italy, provides stimulating and original evidence here, even if the method remains descriptive and too general from the present viewpoint, which requires detailed functional analysis. The case of Swiss and German firm investments in Alsace (Datar, 1974) is another example of beneficial impact, reflecting a logic of minimizing the costs of production at the level of single establishments.

However, the most revealing and decisive evidence of the existence of a spatial and hierarchic division of labour are its consequences for the labour market, consequences which can easily be measured.

Thus a recent synthetic study (Ricq, 1981) covering all of Europe's frontier regions reveals the existence of important differences in wage levels on each side of the frontiers studied. This provides favourable opportunities, and allows room to manoeuvre, for the creation, maintenance or development by firms in the richer region of those segments of production which are strongly influenced by labour costs; these jobs will necessarily be available to workers commuting from the other side of the border.

The duality of the labour market becomes the thermometer of the catalytic role of the frontier - though often for the region on only one side - in the process of spatial diffusion of activities. This situation is clearly illustrated by studies of the border zone between Mexico and the USA (Hansen, 1981; Peach, 1986). It is also present, however, though to a lesser degree, in the case of the frontier triangle of "Alsace-Baden-Basel" and in the axis of the French-Swiss Jura and the Ticino (Switzerland-Italy). In the case of the Ticino (URE, 1981), which has been studied by the national programme of research on "regionalization", the number of border workers increased between 1955 and 1974 from 4,000 to 33,000 (32,000 today). Such employees thus comprise one in every two industrial workers, and one in every five in the entire region. The structure and behaviour of the canton's labour market are strongly characterized by an institutional dualism, the two segments of which are:

- a "free" labour market, comprising the supply of and demand for "home" labour, that is people who can live freely in the canton and change jobs and profession. These are Swiss or foreign workers with a residence permit, or foreigners already admitted to Switzerland on an annual permit (only 7,000 out of 24,000 residents) which allows them a "right" to the renewal of the permit and to professional mobility.
- a "controlled" labour market, comprising the supply of and demand for "external" labour, that is essentially border workers (32,000) and, to a lesser extent, seasonal workers (8,000). The latter are dependent on an administrative authorization for their recruitment and for the renewal of their working contract from one year to the next. Their mobility is thus restrained in the same way.

To this institutional dualism corresponds largely, although not entirely, a socio-economic dualism (Rossi, 1982; Maillat, 1981) related to the characteristics of jobs and workers, namely:

- a "primary" market, whose characteristics are stability of employment and labour, autonomous or even dominant companies, relatively high wages and good working conditions, the existence of rules restricting the employer's behaviour (generally in the form of collective labour agreements), and qualified tasks and jobs allowing a professional career;
- a "secondary" market whose characteristics are the opposite of the primary market, namely instability of jobs and workers, a subordinate position for many companies, low wages and unattractive working conditions, the near absence of collective agreements and hence greater control by employers over their employees, less-skilled work and lack of advancement possibilities.

We shall end the reference to the Ticino case, which seems closely to reflect similar current procedures in other regions, by stressing once more, as in the first part of this chapter, the need for understanding these findings in their historical context. In this particular case, we are talking of the period of liberalization of trade and of economic growth which lasted from the 1950s to 1974. After this date and with a new phase of industrial restructuring, new processes have emerged which are impacting upon the structural framework described above. These are to some extent changing the hierarchical position and the functional role of the production activities of this frontier region.

FRONTIER REGION DEVELOPMENT IN A PERIOD OF INDUSTRIAL AND TECHNOLOGICAL RESTRUCTURING

Since the 1970s, industrial and technological restructuring has generated new dynamisms in the process of spatial diffusion of economic activities in Europe. Research is currently under way on their identification and on the role they play - we are talking, mainly, of the studies of the GREMI - but it is already clear that the trend of discrimination characteristic of the earlier simple spatial division of labour is undergoing unexpected evolutions and more and more complex processes (Aydalot, 1986).

In this new phase of economic development, technological change seems to be associated both with processes of regional repolarization (Camagni, 1986), and with a reversal of earlier spatial hierarchies. Thus in some key sectors, the arrival of new products or the speeding up of the life-cycle phases of existing products demands a repolarization towards urban-industrial centres (old or new), which are the only areas possessing the human and technical capacity to integrate the innovation. On the other hand, some historically "poor" regions are exhibiting new dynamisms, while economic problems have appeared in a number of traditional major industrial regions.

In this evolution of the spatial division of labour, the balance of power between big and small or medium-sized firms appears to have changed considerably in favour of the latter. This has mainly occurred where small and medium-sized enterprises have been able successfully to effect a transition from capacity sub-contracting to production for specialized markets.

Finally, various studies have drawn attention to the possibility of original regional development, with its own internally-generated industrial dynamism. This may involve the phenomenon of spin-offs or the creation of "industrial districts", reflecting different types of division of labour or successful external research and technical development by groups of small enterprises belonging to traditional sectors (Bagnasco, 1985).

What then is the significance of these changes for Europe's frontier regions? Certain inductive hypotheses can be put forward on the basis of the three specificities attributed to frontier regions in the previous section, namely their economic characteristics linked to a proximity effect, their social characteristics linked to the flexibility of the labour supply, and their cultural characteristics related to the degree of permeability of local society.

Generally speaking, in the context of the spatial dynamics of recent technological and economic change, frontier regions do not appear to offer the same particular assets which were of significance in the classical period of the spatial division of labour, and which gave these areas a specific and meaningful place (for better or for worse) in the spatial hierarchy of the distribution of activities. Indeed, from the economic point of view, the proximity effect and the characteristic of being a zone of conflict as well as of contact between national spaces would seem likely to decrease their attractiveness for productive investment under present conditions.

If the spatial division of labour in Europe is becoming more complex, more difficult to interpret, it does not yet mean that there is a trend to strengthen local and territorial co-operation. On the contrary, in terms of innovation mechanisms, large firms seem to be redefining and specializing their functional and spatial organization in a way that goes beyond the logic of proximity and territorial linkage. As far as the innovation process is concerned, there seems on the one hand (Aydalot, 1986) to be developing a phenomenon of "non-spatial internalization" by large corporations, while on the other, firms are also promoting "exteriorization" through increasing contact with specialized research institutions producing knowledge of global relevance.

From the social point of view, this new phase also involves a changing perception of labour requirements. Technological changes are creating, simultaneously, a need for both highly-qualified and less-qualified workers. This means that the problem of the dual segmentation of the labour market - between a primary market and a secondary market - cannot now so easily be resolved by assigning one type of skill to a particular branch unit or establishment. From now on, a "mix" is needed, which does not correspond necessarily to the classic view of the labour market of frontier regions. Above all, it is necessary to adapt local labour market structures (for example those concerned with training) to the rapid contemporary changes in labour demand. Thus there is not only a need for elasticity in quantitative or horizontal terms, but also for elasticity in qualitative, functional and intersectoral terms.

From the cultural point of view, the frontier region characteristic of permeability, with its corollary of fragile and uncertain regional identity, would seem to be unattractive in the context of regional development based on local co-operation and interaction. This contrasts with its positive role in the previous phase of spatial diffusion of corporate activity.

Must we therefore conclude that the traditional thesis stated in the first paragraph and related to the unavoidable handicaps of Europe's frontier regions should be rediscovered and extended?

We do not think so. Rather, we would argue that the theoretical context of the spatial and hierarchical division of labour remains valid, but requires adaptation to the characteristics of the current historical phase (conjunctural and structural). The latter, with territorial and institutional specificities appropriate to each type of frontier region, determine the positive or the negative sign of the discriminant effect of frontier location from the point of view of external or internal spatial organization.

Nevertheless, an initial policy conclusion which may be drawn from this deductive reasoning is that in future, frontier region locational advantages will not result automatically from the region's special position, but in contrast will need actively to be identified and built up.

AN EMPIRICAL TEST: THE CASE OF THE CANTON OF TICINO

An evaluation of the present situation, particularly in terms of the processes of technological change at the level of the frontier region, is possible in the case of the Swiss canton of Ticino, where we have conducted (Ratti and Di Stefano, 1986; Di Stefano, 1986) a co-ordinated survey within the framework of the GREMI. Unfortunately, as far as we know, this is the only example of a frontier region which has so far been studied from this point of view.

In this case, what part have technological changes played in recent regional economic development? And have the technological operations of local firms been hindered by the frontier, as suggested by one of the hypotheses mentioned earlier? The answer is in some ways surprising, and contains features indicating positive territorial characteristics for future development. Full details of the research findings are contained in the references given above. But in summary, the most significant results are:

- after ten years of "restructuring", the region's economic structure still shows the characteristic features of the classical phase of the spatial division of labour;
- the labour market is characterised by a duality which allows certain branches and traditional establishments (clothing, textiles, watches, mechanical engineering) to maintain production based on plentiful and cheap labour, made up of border workers. Notwithstanding marked annual fluctuations (around 15%), overall employment growth has ceased, after reaching what appears to be a spontaneous saturation level of 34,000 workers in 1974;
- half the firms (127 replies) in the region have not planned or not been able to adopt significant innovations in processes or finished products (see Figure 10.2);
- new mechanisms and structures are being superimposed upon the effects and structures of traditional processes of the spatial division of activities. These indicate a change in the nature of spatial integration of part of the Ticinese industrial economy. As a result, one-third of the respondent firms in our survey (response rate 50%) have adopted innovations by imitation, while a further 15% are active innovators (Figure 10.2);
- among these cases, we are specially interested in the importance of the phenomenon of sub-contracting establishments which have succeeded in getting rid of the model of mass-production sub-contracting, and adopted instead the model of specialized sub-contracting. In a region where sub-contracting firms represent an important phenomenon - 153 firms out of 234 participating in the survey - we have found that 64 of

Figure 10.2 Relation between innovation, type of production, market and branches

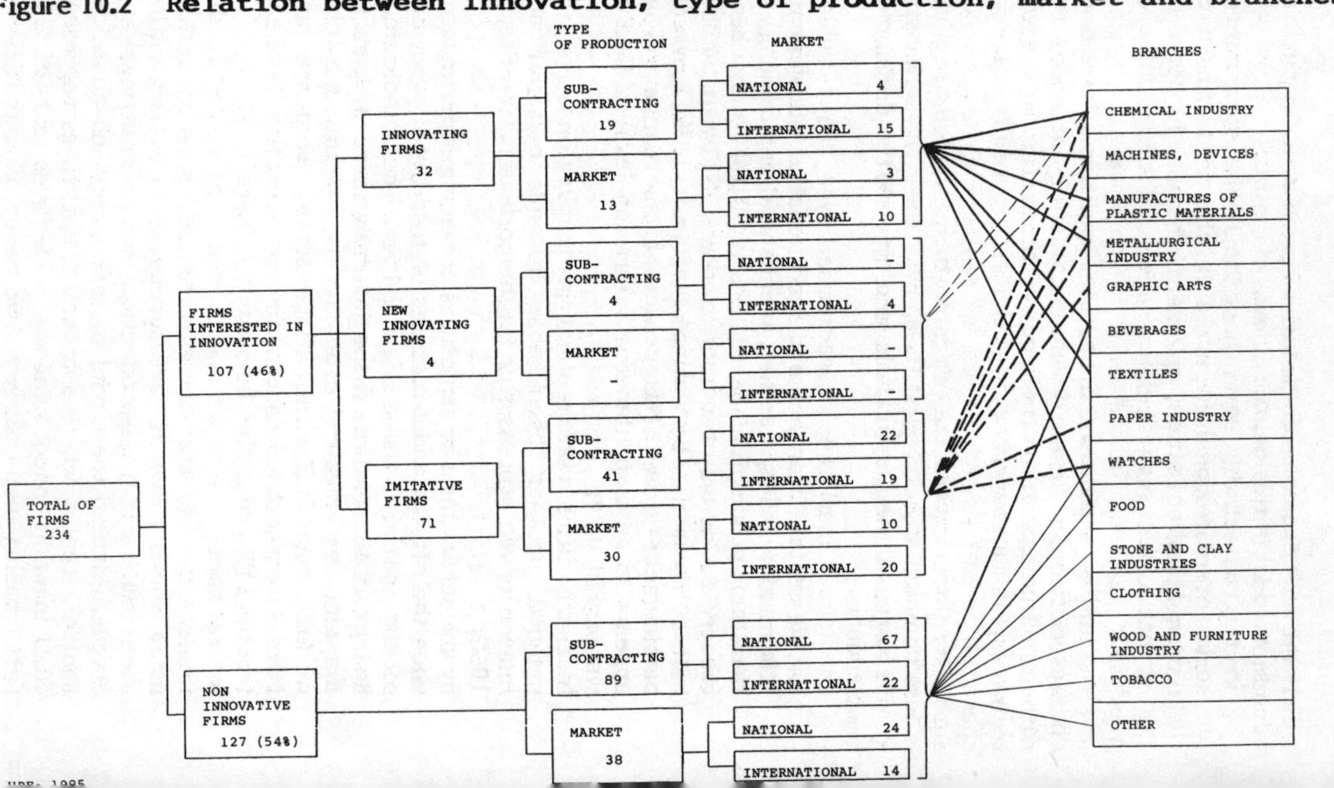

them (42%) could be classified as innovating firms: two thirds are imitative firms, while the other third are among the actively innovating firms. These companies serve international markets (38%) more frequently than national markets (26%), a pattern which is the opposite of that for sub-contracting non-innovative firms.

As far as withdrawal from the traditional model of mass-production or quantity sub-contracting is concerned, with its minimization of the costs of production of certain segments, it is clear that this trend is more frequently characteristic of firms enjoying a relatively high degree of autonomy (local contractors or foreign-owned companies). It is less common in subsidiary Swiss companies which settled in the region during the 1960s and the 1970s looking for border labour;

- another interesting phenomenon concerns the region's industrial dynamism in the form of the creation of new firms. One firm in five, that is to say 46, has been created during only the last ten years (Figure 10.3); and half of these can be classified as innovative firms. A particular feature of this is that the Ticino region appears to have become attractive for foreign companies (from the frontier zone or some other foreign origin) working on commission for an international market. This new situation is certainly not related to the wage factor, as in the traditional case of 1960s' branch units of Swiss companies, but, as revealed by the survey, is a response to the region's geographical situation, its good communications network, and the social and political climate of this Italian-speaking Swiss canton. As these firms generally serve specialized markets, the problem of a lack of suitably qualified workers often arises. In consequence, companies are forced to search for such labour on the other side of the frontier, and this is changing the nature of frontier interaction;
- finally, the research reveals some degree of recent local spin-off of new enterprises (15 cases involving about 500 jobs). These are companies created by one or several entrepreneurs who were formerly employed by another local firm. The most interesting cases (table 10.1) are all linked to a large electronics machinery firm (1,200 jobs), and are the result of two circum-

Figure 10.3 **URE/GREMI - New firms (1975-85) : origin of the new contractors** (types of firms, professions, geographical situation)

	Big firms				Small and middle size firms				Services		Other firm	
	TI	CH	FR	OFO	TI	CH	FR	OFO	TI	OFO	TI	TOT
Contractors	1		1	3	6	2	8	4				25
Engineers	3			1	1							5
Technicians		1			3	2		1	1			8
Administrators				1	1	1						3
Employees/workers					1							1
Miscellaneous	2									1	1	4
	6	1	1	5	12	5	8	5	1	1	1	46
	13				30				3			

TI = Ticino
CH = Rest of Switzerland
FR = Frontier region
OFO = Other foreign origin

46:
10 individual firms
24 firms with several associated founders
12 coming directly from other firms (= spin-off)

TOTAL : TICINO 20 44,0% → the 12 cases of small and middle size firms of Ticino are hardly superior to those of the FR

CH 6 13,0% → in the past the phenomenon was much more important

FR 9 19,5% → new and interesting phenomenon for the frontier region (particularly for the small and middle size firms)

OFO 11 23,5% → new and interesting phenomenon for the frontier region (particularly for the small and middle size firms)

URE/GREMI : A.D. 3.6.1986.

Table 10.1 Spin-off New Firms in Ticino: The Most Interesting Cases

Firm code	Year of creation	Type of activity	Number of jobs	
			At the beginning	1986
1.	1982	Industrial automation	7	20
2.	1985	Industrial automation	1	1
3.	1970	Production of printed circuits	3	10
4.	1981	Production of static continuation groups	47	115
5.	1982	Mechanics software	2	14
6.	1983	Development and connections of terminals and central units	1	5
7.	1978	Industrial automation	2	5
8.	1982	Industrial electronics	4	110
9.	1979	Production of gas turbines and management units	10	30
		Total employed	77	310

Source: URE: 7.86

stances. One is voluntary decentralization by the company, which allowed the departure of the manager and staff responsible for a particular sector of production, even letting them have the machines and the clientele. The other is the departure of staff who had developed a stock of technological knowledge and industrial experience and who have created their own service companies in areas of software and industrial automation.

In drawing general conclusions from this empirical study of phenomena related to technological innovation in a frontier region, two questions will be considered. Firstly, what kind of innovation factors of a territorial type could have acted in the case of recent industrial development in the Ticino region? And secondly, has the frontier played a discriminant role here?

To answer the first question, we must recall the classification of territorial innovation factors suggested by Aydalot (1986, 39), relating to their ability to facilitate one or another type of innovation. These are:

Innovation factors of type 1. Aspects corresponding to the nature of the local industrial network

- number of firms;
- industrial structure (size of the firms, nature of the relations and linkages between firms);
- degree of dependence on external agents;
- sector structure;
- importance of research in local development.

Innovation factors of type 2. Factors of attraction

- Transport, communication and telecommunication facilities;
- Training facilities;
- Quality of life attributes

Innovation factors of type 3. Factors of co-operation

- Research facilities (universities, public laboratories);
- Interpersonal exchanges;
- Information exchanges between the firms;
- Venture-capital availability, enterprise nurseries.

The recent developments observed in the Ticino, an example of a peripheral frontier region with a young but relatively traditional industrial structure, seem to place it in the context of the innovation factors of type 2. In other words, innovative industrial development here is related to factors of attraction, such as good transport, communication and telecommunication facilities (the Gothard axis, between Zürich and Milano; success of domestic air travel), and an attractive quality of life in terms of residential environment. A possible further influence may be Lugano's role as Switzerland's third largest financial centre, with its concentration of high-level international services. The latter have expanded rapidly in recent years, much faster than might have been expected from the region's own, rather modest, development potential. The factors quoted could have improved, at least indirectly, the image of this region in the eyes of manufacturing investors.

On the other hand, factors of type 1, related to the local industrial network, and factors of type 3, related to co-operation elements, seem to operate in a contrary direction, notwithstanding the oft-quoted phenomenon of local spin-offs.

What then of the second question posed above? The answer here is that the frontier seems to have played once again an important positive discriminatory role. Its existence has focussed interest and investment on the Swiss side, thus increasing the attraction factors already quoted. The role of separation filter played by the frontier has acted to channel to the Ticino region activities influenced by broader aspects of Switzerland's image, such as its political and economic stability.

But the frontier has also played a part in its positive function of contact zone. Final easing of frontier constraints on trade in goods and services has allowed local Ticinese industrialists, concerned over problems of restructuring and of national peripherality, to create new links with the Italian border frontier zone of Lombardia. This has helped compensate particularly for local deficiencies in research infrastructure. One third of external research needs are therefore now satisfied in Switzerland, one-third in Lombardia, and one-third elsewhere.

In concluding this section, we should also emphasize how recent industrial development seems to have occurred in spite of the local context, and thanks to factors external to the region. Many of these external factors concern the frontier zone, to the point where the "internationalization" of labour - by a transfrontier policy - and access to R & D services in a neighbouring country, can become important advantages in regional industrial development.

CONCLUSIONS

The problem of frontier regions concerns a considerable number of areas in Europe. However, assessment of frontier region development is often difficult. There is a risk of promoting empirical observations - most often on the penalization effects of the frontier - to the rank of theory, as well as a risk of exaggerating the role of the frontier as a prime factor explaining the regional experience.

In contrast to these views, it has been argued here that it is necessary to interpret the socio-economic and spatial evolution of frontier regions on a higher level, within a more general process of economic and technical development, and to analyse events within the framework of the concept of the international division of labour. The specificities and particular effects of the frontier - with its politico-institutional, social and cultural components - can then be viewed as acting in a discriminatory way, to intensify, diminish or even block processes of wider validity.

Thus frontier space can be conceived as generally existing within a hierarchical and imbalanced spatial development framework. Within this, it can often take on the role - positive or negative - of catalyst of tensions and forces acting inside the socio-economic and politico-institutional subsystems. Thus, for example, phenomena of labour market dualism become particularly evident in the frontier region case.

A dynamic and historical approach (which is unfortunately lacking in specific studies) is fundamental to this analysis. It is essential explicitly to consider different past phases of spatial economic growth, as well as the development of the politico-administrative components of the countries concerned (table 10.2).

Current technological changes are resulting in new economic trends and structural impacts on spatial dynamics. These include repolarization, a reversal in the traditional spatial division of labour, changes in the nature of segments of production, and the emergence of new economic activities. These trends all appear to be less sensitive to frontier effects and discriminatory impacts (table 10.2).

Table 10.2 Spatial Consequences of Economic and Technological Development for an "Open Frontier" Region

Phases		Processes		Consequences
1st case:	Economic phase of polarization (polarization theory)	effects of economic concentration	→	penalization of peripheral zones
	1850-1950 in the case of Switzerland and Italy	effects of politico-administrative concentration	→	effects increasingly penalizing frontier zones
2nd case:	Economic phase of the spatial division of labour (theory of the product cycle; theory of the spatial division of labour) 1955-75 in the case of Switzerland and Italy →	effects of redistribution and spatial diffusion of certain segments of industrial production	→	discriminatory effects of the frontier; positive benefits in the case of frontier regions with high attraction power
3rd case:	Phase of industrial restructuring → 1974 →	new factors: . process of repolarization or reversal in the spatial division of labour . changing nature of the segments of production . new emergent activities (innovating contexts)	→	less evident discriminatory effects of the frontier (+ ; -); positive effects possibly linked to attraction conditions and to specific regional policies

Source: VII.1986/RR

Finally, from the policy perspective, it may be argued that in future, possible territorial attraction factors which frontier regions may be able to offer should not simply be determined by historical circumstances. Rather, they should be deliberately identified, enhanced and extended, not only in economic terms, but also in social, psychological and cultural ones. This approach should involve increased trans-frontier co-operation, as already advocated by some commentators and politicians. Such a policy thus needs only to be re-emphasised and spelt out more fully in concrete terms.

REFERENCES

Aydalot, P. (1986) Les technologies nouvelles et les formes actuelles de la division spatiale du travail, Dossier du Centre Economie, Espace, Environnement, Cahier No. 47, Paris

Bagnasco, A. and Trigilia, C. (1984) Società e politica nelle aree di piccola impresa, Angeli, Milano

Biucchi, B.M. (1959) Le premesse economiche per la difesa dell' italianità, Nuova Società Elvetica, Tip. Pedrazzini, Locarno

Biucchi, B.M. and Gaudard, G. (eds) (1981) Régions frontalières, Georgi, St Saphorin

Bottinelli, T. (1985) Immagini e aspetti di demografia ticinese, L'Almanacco 1986, Bellinzona, pp.151-5

Camagni, R. and Rabellotti, R. (1986) Innovation and territory: the Milan high-tech and innovation field. In P. Aydalot (ed.), Milieux innovateurs en Europe, GREMI, Paris, pp. 101-25

Christaller, W. (1933) Die zentrale orte in süddeutschland, Wissenschaftliche Buchgesellschaft, Darmstadt, reprinted 1980

Datar (1974) Investissements étrangers et aménagement du territoire, Datar, Paris

Di Stefano, A. (1986) Indagine sui processi innovativi in atto nel settore industriale ticinese. In URE, Rapporti semestrali, 1986/1, XII, 57, p. 19

Doeringer, P.B. and Priore, M.J. (1971) Internal labor markets and manpower analysis, Lexington Books, Lexington

Gaudard, G. (1971) Le problème des régions-frontière suisses. In Les régions-frontière et la polarisation urbaine dans la C.E.E., Cahiers de l'I.S.E.A., pp. 3-4

Guichonnet, P. and Raffestin, C. (1974) Géographie des frontières, PUF, Paris
Hansen, N. (1983) International co-operation in border regions: an overview and research agenda, International Regional Science Review, 3, pp. 255-70
Hansen, N. (1981) Mexico's border industry and the international division of labor, Annals of Regional Science, p. 15
Hansen, N. (1977) Border regions: a critique of spatial theory and a European case study, Annals of Regional Science, p. 11
Jeanneret, P. (1985) Régions et frontières internationales, EDES, Neuchâtel
Leimgruber, W. (1984) Die Grenzüberschreitende Region als Verhaltensraum, Habilitationsschrift, Basel (mimeo)
Loesch, A. (1940) Die räumliche Ordnung der Wirtschaft, G. Fischer, Jena
Maillat, D. and Jeanneret, P. (1981) Jura, canton de frontière, Groupe d'études économiques, Neuchâtel
Malecki, E. (1983) Technology and regional development: a survey, International Regional Science Review, 2, pp. 89-125
Michalet, C.A. (1976) Le capitalisme mondial, PUF, Paris
Peach, J. (1986) Social and economic issues concerning frontier markets between USA and Mexico, New Mexico State University, mimeo
Ratti, R. (1971) I traffici internazionali di transito e la regione di Chiasso, Ed. Universitaires, Fribourg
Ratti, R. and Cavadini, A. (1986) Stato e prospettive delle attività di spedizione a Chiasso, Comune di Chiasso
Ratti, R., Bottinelli, T., Cima, T. and Marci, A. (1982) Gli effetti socio-economici della frontiera: il caso del frontalierato nel Cantone Ticino, URE, Quaderno No. 15, Bellinzona
Ratti, R. (1984) Une analyse spatiale d'une activité de service sporadique, URE, Bellinzona
Ratti, R. and Di Stefano, A. (1986) L'innovation technologique au Tessin. In P. Aydalot, (ed.), Milieux innovateurs en Europe, GREMI, Paris, pp. 321-43
Ricq, C. (1981) Les travailleurs frontaliers en Europe, Antrophos, Paris
Rossi, M. (1982) Travailleurs frontaliers, marché du travail et structures economiques: le cas du Tessin, Revue Syndicale Suisse, pp. 7-8
Spehl, H. (1982) Wirkungen der nationalen Grenze auf

Betriebe in peripheren Regionen - dargestellt am Beispiel des Saar-Lor-Lux-Raumes, Universität Trier (mimeo)
Urban, S. (1971) L'intégration économique européenne et l'évolution régionale de part et d'autre du Rhin, Economie et société, 5, pp. 603-35
Vernon, R. (1974) Les entreprises multinationales, Calmann-Lévy, Paris

Chapter 11

The Role of Innovations in Regional Economic Restructuring in Eastern Europe

Ewa Glugiewicz and Bohdan Gruchman

INTRODUCTION

The acceleration of technical progress has become one of the basic issues of economic reforms undertaken in many East European countries in the 1980s. Directions of changes in their economic systems are country specific, but the basic issue is always the same: it is the problem of relatively weak innovative capacity on the part of the national economy, a problem which has become especially visible since the 1970s.

First, let us examine the nature of normative planning in these countries, and of the highly centralized management system of the national economy. Both these approaches, to greater or lesser degree in particular countries, have dominated economic planning throughout Eastern Europe, with the exception of Hungary. Until the early 1980s and even nowadays they were and are visible in most of these countries. These systems incorporate a certain complex of basic and indirect attributes, which are commonly linked and together create conditions for specific features of development and a particular structure of economic activities (Balcerowicz, 1984).

The basic attributes of these systems are the following:

i) an imperative mechanism of planning and a highly centralized system of management which means that the central authority alone has the right to determine the volume of production and to distribute resources;

ii) a hierarchical system of organization with a powerful and highly developed central administration operating through a system of subordination;

iii) the centralization of investment decisions and a

monopolistic position of central state agencies in the field of creation, fusion, deconcentration and structural change as well as in the closing down of enterprises;

iv) a centralized redistribution of the economic surplus, the absence of self-financing of enterprises, and a banking system in the form of a monobank.

The attributes described above are strongly interconnected. In addition, they are linked with other indirect attributes of the economy such as the monopolisation of supply, an essential limitation of the independence of enterprises, a rigid price formation system, imbalanced consumption goods and capital markets, relatively weak linkages with the world market, and poor export competitiveness.

What are the effects of the economic system described above in the field of technical progress and innovation? The correct answer can be found only through micro-economic analysis at the level of the individual enterprise. We are specially interested in examining the external functioning conditions of enterprises within a planned economy. These conditions focus on two elements (Wilczynski, 1984). One comprises the main principles of the functioning system, known as "the rules of the game", the other the economic parameters on which decisions taken in an enterprise are based.

In market economies the principles of the system are relatively stable and the economic parameters are determined by market mechanisms. In planned economies, however, changes in "the rules of the game" are frequent. The economic parameters are determined and dictated by central authorities while the market mechanism is extremely limited. Individual enterprises therefore try to minimize the risk of non-execution of plan imperatives; they seek to obtain minimum planning tasks and, at the same time, to secure as ample a supply of the means of production as possible.

Price formation based on factual costs and administratively-determined prices which do not reflect conditions of supply and demand leads to ineffective decisions and to a considerable waste of resources during the production process: the waste can be included as a production cost in the price of the product. In addition, enterprises generally operate by a series of short-term decisions, most of them having no long-term development strategy.

To sum up, the systemic conditions in most East European countries do not force individual enterprises to act according to micro-economic effectiveness, that is, to undertake objectively rational decisions. Enterprise behaviour of this kind is one of the main causes of low innovation capacity in centrally planned economies. Under existing conditions, innovations which considerably diminish costs or improve the quality of products are not profitable for enterprises. Innovative activities disturb plan execution, while in any case, given imbalanced market relationships between producers and customers, every product, even those which are expensive and of low quality, can be sold. Thus enterprises only have weak motivations for technological changes in their current activities. Greater innovation possibilities occur with new investment, especially when new industrial units are constructed. New establishments with up-to-date technology can play the role of innovation-creating units. The wide-ranging and active investment programmes of most East European countries between 1950 and 1975 resulted in the construction of many industrial units with a relatively high technological level. However, the growing centralization of investment decisions, particularly in Poland, the German Democratic Republic and Czechoslovakia, together with growing capital scarcity, has led since then to an investment policy restricted primarily to the development of only a few consistently high-demand sectors, such as coal mining, energy, metallurgy and chemicals.

In centrally-planned economies the government undertakes innovation policy centrally, aiming at increasing the innovation capacity of the economy. Central authorities treat R&D activities, which are undertaken to a great extent by separate institutions, as a main vehicle of "pumping" technical progress into the economy (Jozefiak, 1984). The general functioning of the R&D sphere is also centrally planned and its activities are powerfully influenced by the central administration.[1] R&D units are frequently separated from enterprises, and are given research tasks and funding directly from central authorities. Being dependent on government subsidies they can act without any special risk. One result of this situation is that R&D activities are only weakly linked with the real demand of enterprises for innovations. This weakness was recognised in the GDR relatively early in the 1970s, when big industrial complexes with their own R&D units were created. Poland and some other East European countries started to change this situation in the 1980s.

It can be seen from the above brief description that the conditions under which enterprises operate, particularly the systemic conditions stemming from the type of planning and highly centralized management, cannot create sufficient motivation for innovative activities at the micro-level within individual enterprises.

In the 1980s this problem of low innovation capacity has given rise to various reforms of the economic system in most East European countries. Although changes already introduced or planned in the economic system are different in different countries, their overall direction is similar: more autonomy for enterprises, less centralized planning, and a greater role for economic efficiency as the leading criterion for economic decisions. The expected results of these changes are also the same: higher innovation capacity of the entire economy.

REGIONAL ECONOMIC RESTRUCTURING

After World War II most East European countries had two main problems to resolve. One was the modernization of their underdeveloped economies, especially in Poland, Bulgaria and Romania. The other was the essential development of backward regions, such as Slovakia, the northern regions of the GDR, and the eastern regions of the USSR. The economic policy of accelerated industrialization adopted at that period as well as a highly centralized system of planning and management did favour the solution of these two main problems. A considerable proportion of the capital resources available was earmarked for investments in selected backward regions. Central decisions gave rise to the establishment of many new industrial enterprises.

The Russian concept of socalled territorial-production complexes provides an excellent example of the developments mentioned above (Bandman, 1976: Aganbegian, 1984). According to this concept, the territorial-production complex is a set of mutually-related industrial enterprises, transport, construction and non-productive units, occupying a certain territory and served by a common infrastructure. Such a complex is formed according to a plan, approved by central authorities. Most TPCs use local mineral resources as a base for leading industries as well as for complementary industries. During the period 1982-5 eleven such complexes were being built in the USSR (Gladyshev and Mozin, 1982). Three of these were in the eastern half of the Russian Federal Republic, three in Kazakhstan and one in Tadzhikistan, that is, in relatively weakly industrialized regions. A good example is the West-Siberian Complex, which is based on oil and gas. Chemical industries - oil refining, synthetic rubber, and plastics - have been developed here, while the manufacture of mining machinery and further expansion of chemical production are planned.

In Poland there are also a few regions corresponding to the Russian territorial-production complexes. These are however of much smaller scale. Good examples are the Konin coal-mining region and Legnica copper district, both of which are in underdeveloped regions. We can also find examples of centrally-planned and managed regional restructuring in Czechoslovakia (in Slovakia), in the GDR (northern and eastern regions), and in Hungary (industrial centres outside Budapest), as well as in Bulgaria and Romania. Most of these regions were developed through extensive investment activity which was not of the territorial-production complex type. However, they all have one thing in common: investment programmes in these regions were devised and implemented according to planning and locational decisions of central authorities. Most of the industrial plants established in the socialist countries of Eastern Europe before the 1980s were sizable units, producing on a large scale, corresponding to technologically-optimum scale characteristics for a given sector. They also employed relatively new technology, for the period concerned. In this way, external innovations were adopted in older regions, a process which often radically changed the regional economic structure. New technologies were mostly created in central R&D institutions, located outside the regions in which the new plants were established.

Thus, the process of regional restructuring was often initiated and accelerated by importing new technology from outside the regions involved on the basis of decisions taken by central authorities. The inadequacy of this form of innovation activity relates to the fact that innovation adoption generally occurred only once, when new plants were constructed. In order to secure continuous application of innovations in new plants it is necessary regularly to import newer technologies from outside or to establish R&D

units in the enterprises concerned. The latter solution initially met many obstacles, particularly of an organizational kind. For example, in Poland in the 1960s, a number of enterprise-based R&D units were closed down, and their activities transferred to industrial boards and to ministries supervising particular industrial sectors. Again, in the GDR during the 1970s, official policy began to favour the creation of large industrial complexes containing many enterprises (Kombinate), and these took over the R&D units previously located at individual enterprises (Harmann and Kaergel, 1985).

This chapter has focussed so far on the modernization of underdeveloped regions in Eastern Europe. However, the model of centrally guided restructuring of the economy applies also to developed regions and to existing industrial centres. Official policy originally intended that these regions should develop less intensively than backward ones. However, it proved impossible to stop or to markedly slow down their development. In practice many industrial centres have increased their economic potential and substantially changed their structure. Good examples of such industrial growth centres can be found in the GDR at Leipzig, in Czechoslovakia at Bratislava, and in Poland at Cracow (Haranczyk, 1986). During the 1960-80 period, industrial employment in Bratislava and Cracow grew by 41% and 43% respectively. And industrial production in these two cities grew much faster than this over the same period. Rapid output growth also occurred in Leipzig, although industrial employment here decreased by 26% over the 20-year period.

The "unplanned" development of old large industrial centres has in some cases been influenced by a centrally-determined deglomeration policy. In these cases, some existing plants have been transferred to ex-urban locations, while the construction of new plants has been forbidden within the city. This was the case in the 1960s in Warsaw, Lodz, Cracow and Poznań in Poland, and in Budapest in Hungary: some limitations are indeed still in force in the last of these cities.

Although the development of large industrial cities was centrally controlled and guided, the construction of new plants or modernization of existing ones was often based on new technologies which had been developed locally. This was possible because of the existence of local scientific research centres and R&D institutions. Hence, development in such cities could be ascribed to local technological innovation.

The restructuring of regions based on their own innovations is usually not a very rapid or spectacular process. It requires special conditions and circumstances if it is to bear fruit. Since such conditions exist only in a few regions, this type of regional restructuring has not so far dominated regional development in most East European countries. However, this situation is now changing radically. Traditional initial industrialization policies for backward regions are no longer effective. New policies of regional restructuring based on internal innovation, or development from below, have to be adopted. That is why we now turn to the analysis of conditions favouring local synergies in East European countries.

REGIONAL PLANNING AND THE FUNCTIONING OF LOCAL SYNERGIES

In all East European countries, national economic growth and particularly industrial development have been strongly dependent on central, sectoral management. However, the level of sectoral dependence has differed as between different countries. The sectoral system of planning and management has developed specially in the Soviet Union, where industries in a given region have been dependent on federal ministries, ministries of a mixed federal and republic status, as well as on a particular republic's own ministries. The GDR system is also relatively centralized. Here most industrial enterprises belong to large complexes (Kombinate), which are supervised by central, sectoral ministries. Even local industry is organized in the form of such complexes. The level of centralization is also relatively high in Czechoslovakia, Bulgaria and Romania. On the other hand, industry in Hungary and, since 1982, also Poland is relatively less dependent on sectoral ministries.

For a given enterprise, a high level of sectoral dependence involving mainly vertical and hierarchical relationships represents a considerable obstacle to horizontal links within a given region or centre. In order to overcome this problem, different organizational and planning instruments are used at the regional level. Co-ordination of horizontal links within the regional economy is mostly implemented through a system of five-year plans. On the one hand, in all East European countries the central plan is accompanied by specific regional profiles. These are based on official administrative divisions, such as the republics in the Soviet Union, Bezirke in the GDR, and voievodships in Poland.

The regional profiles of the central plan constitute a base for the co-ordination of sectoral plans in a given region. On the other hand, regional authorities have the power to co-ordinate sectoral activities through planning of manpower, transportation and construction. They are also authorized to prepare regional balances of certain minerals, raw materials and products as well as balances of population money incomes and expenditures. They are required to prepare five-year plans of overall regional development covering all aspects of the economy, regardless of whether they are centrally or regionally guided. Of course, regional authorities are also obliged to control the execution of these comprehensive plans.

As can be seen from the above short description, formal possibilities exist for the horizontal co-ordination of the entire regional economy in spite of its sectoral dependence. However, in practice the elaboration of complex plans and determination of regional balances, as well as their execution, encounters many difficulties. A typical list of these difficulties is given by Vorontsoff (1984: see also Tampiza, 1985) for the USSR. They include the development of inconsistencies between sectoral and regional plans, the very low quality of many regional plans, and frequently-considerable delays in their elaboration. Very similar weaknesses have been observed and criticized in other socialist countries. For example, there are many difficulties in the co-ordination and execution of territorial plans in the GDR despite the fact that the local authorities there are in a position to use a relatively large number of co-ordination instruments (Lange, 1986).

Horizontal co-ordination of the regional economy is still more difficult within short-term - one year - planning. Operational, sectoral management dominates over horizontal aspects and relations. With respect to plan execution, local and regional authorities can effectively control only those spheres of the regional economy which are administratively supervised by local and regional authorities. Traditionally, the regionally or locally-dependent sphere of the economy covers only small-scale industries and services, agriculture, local transport, retail trade, housing, communal utilities, and other services of local importance such as health care and primary and secondary education (Opallo and Marckinkowska-Kedzierska,

1986). It must be stressed, however, that all these spheres are doubly-subordinated, in that they are both regionally or locally-dependent, and centrally-dependent, at the same time. Regional and local authorities are therefore heavily restricted in their actions.

What influence can local and regional authorities exert on innovation creation and technological change in their regions? A major constraint here is that regional authorities cannot directly influence the activity of research centres and R&D institutions. In most East European countries, research centres are directly subordinated to the Ministry of Research, whereas R&D institutions are responsible to different industrial branch ministries.

All this does not help very much in developing local synergies with respect to technological innovation. However, in spite of this there are some positive examples which show that regional authorities can be very active in promoting co-operation between industrial plants and local R&D institutions. This is the case for example in the GDR, where regional authorities are responsible for "intensification" of development factors within their territory. They are legally obliged to co-ordinate the activity of enterprises and R&D institutions (Bonisch, Moks and Ostwald, 1982). In the Soviet Union there are examples of such co-operation in large cities, particularly in Moscow, Leningrad, Novosibirsk and certain other Siberian cities where there are seats of the Siberian Branch of the Soviet Academy of Sciences (Ed. "Nauka", 1978).

It can be argued that the regional innovation process in industry, particularly in small and medium-size enterprises, could be much more intensive if such enterprises and regional authorities had greater autonomy. Such a situation with regard to the economic and planning system does exist nowadays in Hungary and to a lesser extent also in Poland. In Hungary, no formal co-ordination of regional and central plans is required. The relations between regional authorities and enterprises are mainly regulated through different economic instruments. The creation of stable sources of financial income for regional authorities is regarded as the most important task for the near future.[2] Similar changes regulating the actions and responsibilities of regional authorities were undertaken in Poland in 1984. More recently, new regulations creating favourable financial conditions and tax-reductions for small-scale regionally-subordinated industries have also been introduced.[3] All these actions should help regional and local authorities to stimulate innovations and through them a faster development and reconstruction of their regions.[4]

Last but not least as a factor in stimulating the local and regional innovation process is the issue of greater financial autonomy for individual enterprises, particularly with regard to finance for future development. In some East European countries the financial system under which enterprises operate has been already changed so as to ensure the autonomy of individual enterprises over financial decisions: in other countries, such changes are planned for the near future (Nowe Drogi, 1985). If such trends continue, decisions concerning regional industrial development will be further decentralized. In this way the creation of favourable conditions for regional development through technological innovation will rest to a much greater extent in the hands of autonomous enterprises and local and regional authorities.

CONCLUSIONS

All the socialist countries of East Europe are now looking for new ways to accelerate economic development. These efforts are being pursued with greater or lesser intensity, and also vary in approach, as between different countries. The most crucial objective in all cases is to increase the innovation capacity of the national economy to its highest possible level. Most governments are committed to continuing and expanding traditional central innovation policy, in the form of "pumping" innovations into the economy from above. But at the same time, most of them are searching for new ways of stimulating autonomous enterprises to create and adopt innovations for themselves.

In this latter context two different approaches are already evident: the Hungarian one which stresses the role of individual enterprises in innovation, and the GDR approach which emphasizes the key role of industrial complexes (Kombinate) in technological change. Thus far, experiences vary considerably so that it is too early to draw any conclusions as to which model will be more effective. It may well be that a combination of both will best fit the conditions of a given country. It should however be noted that each policy approach has specific spatial effects, and creates different conditions for innovation adoption, diffusion and technological change as a basis for restructuring the regional economy.

NOTES

1. New regulations concerning R&D activity in Poland are a good example here. See Fronczak (1986) for further details.
2. This is based on information in an official document entitled Directions of change in the management system in Hungary, produced by the Planning Office of Hungary, Budapest, 1985 (Russian language translation).
3. These are listed in the State Council and Council of Ministers order concerning the development of small state enterprises, published in Monitor Polski, 31 May 1986, no. 4.
4. A further sphere of regional intervention which is not considered here is development at the micro-regional level, which is a very important element of decentralisation policy in socialist countries: see Gruchman (1982).

REFERENCES

Aganbegian, A.G. (ed.) (1984) Territorial-Production Complexes: Planning and Control. In Russian, Isdatelstvo "Nauka", Sibirskoie Otdeleniye, Novosibirsk

Balcerowicz, L. (1984) System gospodarczy a techniczna innowacyjnosc kraju (Economic system and innovation capacity of the national economy). In Innowacje w gospodarce, PWE, Bialystok-Bialowieza

Bandman, M.K. (ed.) (1976) General questions of modelling territorial-production complexes. USSR Academy of Sciences, Novosibirsk

Bonisch, R., Moks, G., and Ostwald, W. (1982) Territorialplannung, Verlag "Die Wirtschaft", Berlin. See particularly section 4.2.2: Wissenschaft und Technik als Hauptfator für die Dynamik und Intensivierung der territorialen Organisation der Produktion, pp. 133-9

Fronczak, K. (1986) Plan postepu (plan of progress), Zycie Gospodarcze, 20, May

Gladyshev, A.A. and Mozin, W.P. (1982) Territorialno-proisvodstvennyie Komplieksy SSSR (Territorial-production complexes in USSR). Politizdat, Moscow

Gruchman, B. (1982) Local and regional development in East Europe: experiences, main issues and perspectives. Regional Development Dialogue, 3, 1, United Nations Centre for Regional Development, Nagoya, Japan, pp. 28-44

Haranczyk, A. (1986) The Comparative Analysis of the Social Development of Cracow, Bratislava and Leipzig in the period 1960-1985. Faculteit der Economische Wetenschappen, Katholieke Hogeschool Tilburg, Reeks "Ter Discussie", 36.10, p.4

Harmann, K. and Kaergel, S. (1985) Koncentracja i kombinatowa organizacja produkcji w przemysle NRD - podstawa wyzszej efektywnosci (Concentrated production organisation in the GDR industry - a basis for higher effectiveness). In Nowe Drogi, Rozwoj i funkcjonowanie gospodarki krajow socjalistycznych, Supplement, November

Jozefiak, C. (1984) Problemy centralnego sterowania innowacjami (Problems of central innovation policy). In Innowacje w gospodarce, PWE, Bialystok-Bialowieza

Lange, U. (1986) Zur Problematik der Produktions-statten, Wirtschaftswissenschaft, 2, pp. 200-8

Ed. "Nauka" (1978) Razvitiye narodnovo Khosiayoistva Sibirii (Economic development of Siberia). Collective work, Ed. "Nauka", Sibirskoie Otdeleniye, Novosibirsk. See particularly chapter 3, Nautchnotiekhnitcheskiy progriess i intiensiffikatsiya ekonomiki Sibirii (Technological progress and intensive development of Siberia), pp. 128-42

Nowe Drogi (1985) Rozwoj i funkcjonowanie gospodarki krajow socjalistycznych (Development and functioning of the socialist countries' economies), Nowe Drogi, Supplement, November

Opallo, M. and Marcinkowska-Kedzierska, M. (1986) Planowanie terytorialne w wybranych krajach socjalistyczneh (Territorial planning in selected socialist countries). In Instytut Gospodarki Narodowej (ed.), Studia i Materiaky, Warszawa, p.102

Tampiza, K. (1985) Sovyeti i komplieksnoye razvitiye territorii (People's councils and comprehensive territorial development), Planovoie Khosiaistvo, 4, pp. 106-9

Vorontsoff, V. (1984) Voprosy soverchenstvovaniya territorialnovo planirovaniya (Problems of territorial planning improvement), Planovoie Khosiaistvo, 1

Wilczynski, W. (1984) Sklonnosc do innowacji a system ekonomiczny (Innovation capacity and economic system). In Innowacje w gospodarce, PWE, Bialystok-Bialowieza

INDEX